Springer Tracts in Civil Engineering

Series Editors

Sheng-Hong Chen, School of Water Resources and Hydropower Engineering, Wuhan University, Wuhan, China

Marco di Prisco, Politecnico di Milano, Milano, Italy

Ioannis Vayas, Institute of Steel Structures, National Technical University of Athens, Athens, Greece

Springer Tracts in Civil Engineering (STCE) publishes the latest developments in Civil Engineering - quickly, informally and in top quality. The series scope includes monographs, professional books, graduate textbooks and edited volumes, as well as outstanding PhD theses. Its goal is to cover all the main branches of civil engineering, both theoretical and applied, including:

- Construction and Structural Mechanics
- Building Materials
- Concrete, Steel and Timber Structures
- Geotechnical Engineering
- Earthquake Engineering
- Coastal Engineering; Ocean and Offshore Engineering
- Hydraulics, Hydrology and Water Resources Engineering
- Environmental Engineering and Sustainability
- Structural Health and Monitoring
- Surveying and Geographical Information Systems
- Heating, Ventilation and Air Conditioning (HVAC)
- Transportation and Traffic
- Risk Analysis
- Safety and Security

Indexed by Scopus

To submit a proposal or request further information, please contact:

Pierpaolo Riva at Pierpaolo.Riva@springer.com (Europe and Americas) Wayne Hu at wayne.hu@springer.com (China)

More information about this series at https://link.springer.com/bookseries/15088

Boris Azinović · Vojko Kilar · David Koren

Assessment of Energy-Efficient Building Details for Seismic Regions

Boris Azinović
Department of Structures
ZAG Ljubljana
Ljubljana, Slovenia

Vojko Kilar
Faculty of Architecture
University of Ljubljana
Ljubljana, Slovenia

David Koren
Faculty of Architecture
University of Ljubljana
Ljubljana, Slovenia

ISSN 2366-259X ISSN 2366-2603 (electronic)
Springer Tracts in Civil Engineering
ISBN 978-3-030-97555-5 ISBN 978-3-030-97556-2 (eBook)
https://doi.org/10.1007/978-3-030-97556-2

© The Editor(s) (if applicable) and The Author(s) 2022. This book is an open access publication.
Open Access This book is licensed under the terms of the Creative Commons Attribution 4.0 International License (http://creativecommons.org/licenses/by/4.0/), which permits use, sharing, adaptation, distribution and reproduction in any medium or format, as long as you give appropriate credit to the original author(s) and the source, provide a link to the Creative Commons license and indicate if changes were made.

The images or other third party material in this book are included in the book's Creative Commons license, unless indicated otherwise in a credit line to the material. If material is not included in the book's Creative Commons license and your intended use is not permitted by statutory regulation or exceeds the permitted use, you will need to obtain permission directly from the copyright holder.

The use of general descriptive names, registered names, trademarks, service marks, etc. in this publication does not imply, even in the absence of a specific statement, that such names are exempt from the relevant protective laws and regulations and therefore free for general use.

The publisher, the authors and the editors are safe to assume that the advice and information in this book are believed to be true and accurate at the date of publication. Neither the publisher nor the authors or the editors give a warranty, expressed or implied, with respect to the material contained herein or for any errors or omissions that may have been made. The publisher remains neutral with regard to jurisdictional claims in published maps and institutional affiliations.

This Springer imprint is published by the registered company Springer Nature Switzerland AG
The registered company address is: Gewerbestrasse 11, 6330 Cham, Switzerland

Preface

The monograph is intended for a wide range of experts included in the design of modern energy-efficient buildings. Scientific and expert features in the monograph help to better understand the seismic risk of energy-efficient buildings and their structural details and awareness of the consequences that may occur due to the unsuitable design of such buildings. The research presented in this monograph is a full examination of the selected critical structural details of energy-efficient buildings, which includes evaluation from various aspects, particularly from the technical and structural and environmental and energy efficiency aspects. The methodology of evaluating the structural details of the building envelope is presented in the monograph on the basis of previously conducted scientific studies on critical structural details and recommendations from standards for energy-efficient and earthquake-resistant construction.

The main objectives of the monograph are: (i.) to contribute to a more detailed knowledge of constructing energy-efficient buildings in earthquake-prone areas; (ii.) to review, evaluate, and analyse the typical details of the thermal envelope of energy-efficient buildings from the aspect of earthquake resistance; (iii.) to raise awareness of the importance of structural details, their impact on the earthquake resistance of buildings, on the prevention of thermal bridges, and the provision of thermal comfort for users; (iv.) to recognise and eliminate unsuitable structural details from the aspect of earthquake resistance in the construction of new buildings and renovations; (v.) to present a methodology for the assessment of building details and examples of good practice; (vi.) to facilitate progress in earthquake engineering and architecture by designing satisfactory earthquake-resistant and energy-efficient details; and (vii.) to encourage cooperation across all professions included in the design of energy-efficient buildings, particularly in the initial phase of the structural design.

The monograph includes the presentation of the most important results of research on earthquake resistance of low-energy buildings carried out by the authors over the past decade and published in various scientific articles. It presents a new methodology for structural detail evaluation from the aspect of their earthquake resistance and energy efficiency. Its purpose is to distinguish between good and bad detail solutions, which must be part of the building design. The final section of the monograph includes the description of "ready to use or not to use" various examples of most frequently used foundation, basement, balcony, and roof details (a total of more than 20, which are described, analysed, and commented). The results of the environmental and energy efficiency and technical and structural parameter evaluation are provided for each detail. These examples can be used to directly transfer knowledge into practice, which already uses some energy-efficient details with unsatisfactory static and seismic resistance.

Ljubljana, Slovenia

Boris Azinović
Vojko Kilar
David Koren

Acknowledgements The author Boris Azinović gratefully acknowledges the financial support for the project InnoCrossLam, which is supported under the umbrella of ERA-NET Cofund ForestValue by MIZŠ and has received funding from the European Union's Horizon 2020 research and innovation programme under grant agreement No. 773324.

All the authors gratefully acknowledge the financial support received from the Slovenian Research Agency (Research Core Funding No. P5-0068 and P4-0430).

The authors would like to thank Laura Cuder Turk for her assistance in editing and proofreading this book.

Contents

1	**Introduction**	1
	References	5
2	**Design of the Thermal Insulation Envelope of Energy-Efficient Buildings**	7
	2.1 »L« Type Structural Assembly Connections	10
	2.2 »T« Type Structural Assembly Connections	13
	2.3 »+« Type Structural Assembly Connections	13
	References	15
3	**Structural Details in Energy-Efficient Buildings**	17
	3.1 Foundations on the Thermal Insulation Layer	17
	3.1.1 Possible Solutions for Foundations on Thermal Insulation	17
	3.1.2 Specifics of Designing Thermally Insulated Foundation Slabs in Non-earthquake-Prone Areas	28
	3.1.3 Influence of the Flexible Layer of Thermal Insulation on the Seismic Response of a Building	34
	3.2 Base Insulation Blocks for Preventing Thermal Bridges in Walls in Contact with Cold Elements	38
	3.2.1 Types of Base Insulation Blocks	38
	3.2.2 Influence of Base Insulation Blocks on Better Environmental and Energy-Efficiency Parameters	39
	3.2.3 Base Insulation Blocks as Parts of Masonry Structures for Bearing Vertical Static Loads	44
	3.3 Details for Preventing Thermal Bridges in Cantilever Structures	45
	3.3.1 Precast Element Solutions for Thermal Bridges in Cantilever Structures	45
	3.3.2 Comparison of the Environmental and Energy-Efficiency Parameters of an Insulated and Uninsulated Reinforced Concrete Cantilever Slab	47

		3.3.3	Selection Principles of Precast Load-Bearing Thermal Insulation Elements Subject to Vertical Static Loads	49
	3.4	Building Connection Detail Between the Roof and Outer Wall		54
		3.4.1	Environmental and Energy-Efficiency Aspects of the Building Connection Detail Between the Roof and an Outer Wall	56
	References			58
4	**Evaluation of Critical Structural Assemblies**			**65**
	4.1	Detail Evaluation Methodology for Energy-Efficient Buildings		65
		4.1.1	Importance of Environmental and Energy-Efficiency Parameters	67
		4.1.2	Importance of Technical and Structural Parameters	68
		4.1.3	Importance of External Parameters	70
	4.2	Environmental and Energy-Efficiency Parameters		71
		4.2.1	Thermal Transmittance of Structural Assemblies (E1)	71
		4.2.2	Continuity of Thermal Insulation (E2)	72
		4.2.3	Condensation and Thermal Comfort (E3)	73
		4.2.4	Influence on the Use of Energy (E4)	75
		4.2.5	Airtightness (E5)	76
		4.2.6	Life Cycle Assessment (E6)	78
		4.2.7	Durability and Sustainability (E7)	80
	4.3	Technical and Structural Parameters		80
		4.3.1	Load-Bearing Capacity (K1)	80
		4.3.2	Minimum Dimensions and Stiffness (K2)	82
		4.3.3	Symmetry (K3)	83
		4.3.4	Continuity or Uniformity of the Load-Bearing Structure (K4)	84
		4.3.5	Eccentricity or a Shift in the Structure According to the Primary Load-Bearing Axis (K5)	85
		4.3.6	Capacity Design Method (K6)	86
		4.3.7	Connections of Primary and Secondary (Non-) Load-Bearing Elements (K7)	87
	4.4	External Parameters		88
		4.4.1	Location (Z1)	89
		4.4.2	Importance of a Building (Z2)	89
		4.4.3	Influence on the Global Analysis (Z3)	90
		4.4.4	Complexity of Construction (Z4)	91
		4.4.5	Penetrations and Openings (Z5)	91
		4.4.6	Economic Aspect (Z6)	92
	4.5	Assessment Based on Evaluation Parameters, Weighting Factors and External Parameters		93
	4.6	Presentation of the Results		98
	References			102

5	**Case Study: Using Methodology to Assess the Selected Details**	107
	5.1 Building Connection Detail Between an Outer Wall and the Foundation Slab	109
	5.2 Building Connection Detail Between the Load-Bearing Balcony Structure and an Outer Wall	117
	5.3 Building Connection Detail Between an Outer Wall and the Unheated Basement	123
	5.4 Building Connection Detail Between an Outer Wall and the Roof	128
	References	133
6	**Conclusions**	135
	References	139

Appendix: Examples of the Use of the Methodology for Evaluating Structural Details 141

Chapter 1
Introduction

The content of the monograph contributes to the attainment of sustainable development goals as part of United Nations Agenda 2030 by considering the structural details on the thermal envelope of energy-efficient buildings (UN (United Nations) 2015). The research performed is in line with the EU Green Deal (EU Commission 2019b) and also targets the development of the new European Bauhaus initiative (EU Commission 2021). A structurally safe, sustainable and energy-efficient details will improve the competitiveness of energy-efficient buildings as a change agent towards a circular and sustainable building sector. As the market for energy efficient buildings spread to earthquake-prone areas, the structural details on the building envelope must be adapted to ensure sufficient structural resistance. Furthermore, with the Renovation Wave in Europe aiming to renovate and refurbish 30–40 million homes in the next ten years, the contribution of this monograph is even more important.

The overall goal of this work is to assess energy-efficient building details for seismic regions with properties meeting the needs and demands of a sustainable-oriented market. The presented results impact several different aspects, which correspond to the attainment of sustainable development goals and also relate to the EU Green Deal:

- **Good Health and Well-Being (SDG3)**: Well-designed building structural details are aimed at increasing the good indoor environmental quality in energy-efficient buildings (e.g. by reducing moisture defects, using materials with less volatile organic compound emissions etc.), resulting in high living comfort which positively affects the health, well-being, and work performance of users.
- **Decent Work and Economic Growth (SDG8)**: Off-site manufacturing of building components, simplicity and automation in the construction of energy-efficient buildings shortens the building process. By implementing well-designed energy-efficient structural details aimed also at modular construction, the safety in the working environment will significantly increase. Well-designed structural details and new concepts can open new markets and hence reach stable economic growth.

- **Industry, Innovation, and Infrastructure (SDG9)**: Development of energy-efficient buildings with innovative structural details brought new products on the market (e.g. load-bearing materials with thermal insulation capability), and raised the potential for industry and market development in the construction sector, especially due to flexibility, adaptability, low-maintenance need and recyclability.
- **Sustainable Cities and Communities (SDG11)**: The next-generation of energy-efficient buildings provides potential for new quality living areas, which can consequently result in more sustainable ways of living. Technological innovation in various forms and digitalisation will contribute to the sustainable transformation of the built environment on all levels (building, district, city).
- **Responsible Consumption and Production (SDG12)**: for the built environment to become socio-economically effective and sustainable, circular resource flows and the circular economy are major driving forces. The work is aimed at the building sector by improving building efficiency (efficient use of natural resources), increasing simplicity of structural details and technology transformation.
- **Climate Action (SDG13)**: The manufacture, transportation, installation, maintenance, and disposal of building products/materials amount to around 11% of global emissions. The methodology in this work is aimed at reducing greenhouse gas emissions and emission intensity in energy use, which is critical considering the urgency of climate action and a priority of the 'greener, carbon free Europe'.

In the monograph, the term 'energy-efficient building' applies to all buildings that take into account modern requirements for a better thermal envelope and a lower energy consumption. Directive 2010/31/EU on the energy performance of buildings (DEUS 2010/31/EU) was adopted at the EU level in 2011 to reduce energy consumption in households. The Directive stipulates the construction of nearly zero-energy buildings beyond 2021, which has so far been prescribed by EU member states in various incentives and acts with mixed success (EU Commission 2019a). The term 'nearly zero-energy buildings' is defined in the Directive and signifies a building with a very low energy consumption, which can cover all losses from renewable sources. The term 'energy efficiency' defines the efficiency of a building in terms of energy used for heating and cooling, ventilation, air conditioning, and efficient hot water supply.

The German passive house standard (PassivHaus; hereinafter referred to as 'the PH standard') can be stated as an example of the building energy efficiency standard. The provisions and requirements of the PH standard have been amended for the design of the first passive house to the present, and the details of passive houses will continue to be developed and improved (Passivhaus Institut 2012). The PH standard is among the first standards to numerically define requirements to achieve high energy efficiency, setting an example for many other building energy efficiency standards. According to Dequaire (2012), standards applied by various countries to regulate the construction of energy-efficient buildings includes Swiss Minergie-P (2009), French BBC-effinergie (2012), Norwegian prNS 3701 (2011), Danish Damnarks Lavenergibygning klasse 1 (2009), and British Code for Sustainable Homes level 6 (2009).

The provisions of all these standards contribute to the understanding of modern energy-efficient building design. The objective of such a design is to build buildings of the highest quality possible by taking into account the local climate, making use of renewable sources and ambient energy (heat and cold), and using locally available materials with a low environmental impact for construction (see, for example, Krainer (2011), Küçüktopcu and Cemek (2018), AzariJafari et al. (2021), Hoxha et al. (2017)). To attain these objectives, the architectural design of buildings is crucial. From the aspect of energy efficiency, it must provide: *(i) a carefully planned orientation and location of a building to optimise requirements for energy conservation; (ii) a favourable ratio between the building envelope and volume to avoid unnecessary division of buildings and elements that could cause thermal bridges; (iii) the sunlight exposure of the building envelope with the thermal energy function; (iv) the shape and ratio of glazing to gain as much heat in winter as possible and protect against excessive sunlight exposure and heating in summer; (v) well-designed and planned surfaces suitable for solar thermal collectors and other devices that utilise solar radiation; and (vi) a good thermal insulation of the thermal envelope and other transparent elements (e.g. windows) on the building envelope.*

Due to all the above-mentioned regulations, the construction of various types of energy-efficient houses has become widespread in Europe in recent years. Common characteristics of such buildings are that the thermal envelope must be uninterrupted—also under the building or its foundations, and thermal bridges must be prevented. Numerous details are used in practice for this purpose, which were developed in Western Europe or Scandinavian countries with a low risk of considerable seismic loads. The structural systems of passive houses were also developed in low seismic hazard areas, making them suitable for vertical and wind loads. However, there is no guarantee that their behaviour will be appropriate and ductile in the event of cyclic seismic loads. It must be noted that inserting soft insulation layers with better vertical load-bearing capacity extends the fundamental period of a structure, as a structure on the thermal insulation layer vibrates more slowly than on the ground. Most such buildings are low-rise buildings and have very short fundamental periods. By extending the fundamental period, they can be moved to the resonance part of the design response spectrum (the periods within constant acceleration plateau). This means that in stiff buildings on poor soil 2 to 3.5-times higher forces acting on the building could be expected. The results of seismic analyses exposed that the height of such buildings must be limited to three or four storeys, depending on the slenderness, stiffness, and mass of the building.

Additionally, so-called base insulation blocks are frequently used in passive houses, which are inserted between the upper part of the structure and the foundation slab or unheated basement to prevent thermal bridges. In an earthquake, the shear strength of walls and columns is very important and may be reduced to the point where the structure's earthquake resistance is no longer sufficient by inserting such insulation load-bearing elements designed only for vertical loads. Studies have shown that such technology transfer can be dangerous, particularly in larger and heavier multi-storey buildings in high seismic hazard areas. Studies have also shown that, when designing earthquake-resistant details, sufficient attention must be paid to their resistance to horizontal actions. The energy efficiency of certain details is inversely proportional to their earthquake resistance. Therefore, we have developed a special detail evaluation methodology, which is presented in more detail in Chap. 5 of this monograph.

Due to the unique character of energy-efficient buildings, it is frequently difficult to meet all the requirements for earthquake-resistant construction, as the environmental and energy-efficiency criteria must also be taken into account to reduce environmental impact and increase thermal comfort for users of buildings. We have found that certain requirements of energy-efficient and earthquake-resistant construction are diametrically opposite. The use of mere technical guidelines provided by standards for earthquake-resistant construction, e.g. Eurocode 8 (CEN 2005), can result in the significant deterioration of details in terms of energy. Such requirements include the uniformity and continuity of the load-bearing structure or structural regularity by height without major changes in its load-bearing capacity and stiffness. These general principles of standards may result in the interruption or deterioration in the thermal envelope, causing thermal bridges in the thermal envelope, which must be prevented to comply with the current energy-efficient construction regulations. On the other hand, the problem can be viewed from the opposite side. Taking into account the recommendations in energy-efficient construction standards on continuous thermal insulation may lead to interruptions and discontinuity in the load-bearing structure, which may significantly change the response in earthquake-prone areas. An interesting example of a problem in this field is fixing balcony cantilevers without a thermal bridge, since thermal insulation is most necessary at the precise location of the highest bending moment reaction of the cantilever.

The structural safety of energy-efficient buildings has not yet been extensively studied, since most first energy-efficient buildings were smaller, making them less vulnerable to seismic actions. The technology of energy-efficient building construction has expanded to almost all types of buildings (including large buildings, such as blocks of flats, large office buildings, etc.). Therefore, higher loads on crucial structural details may be expected. In the monograph, we analyse as many details of energy-efficient buildings as possible, and find their specifics and limitations of their use in earthquake-prone areas. The most important general findings of our research in this field are: *(i.) the principles of energy-efficient buildings and their structural details for preventing thermal bridges can reduce earthquake resistance of structures in comparison with conventional earthquake-resistant construction; (ii.) certain details of energy-efficient buildings used in non-earthquake-prone areas cannot be transferred to earthquake-prone areas; (iii.) suitable design can improve the response of energy-efficient buildings for them to comply with current regulations on the design of structures for earthquake resistance and the principles of energy efficiency; (iv.) with the new proposed methodology, we can establish the extent to which environmental and energy-efficiency, and technical and structural aspects are taken into account in structural details, and recognise critical (poorer) details in terms of earthquake resistance.*

In the monograph, the possibilities to use the concepts and details of energy-efficient buildings in earthquake-prone areas are explored, and analytical and practical approaches to improve their safety are presented. The authors aim to facilitate progress in earthquake engineering, architecture, and other disciplines dealing with the design of quality earthquake-resistant and energy-efficient details of buildings.

References

Azarijafari H, Guest G, Kirchain R, Gregory J, Amor B (2021) Towards comparable environmental product declarations of construction materials: Insights from a probabilistic comparative LCA approach. Build Environ 190:107542

CEN (2005). Eurocode 8: design of structures for earthquake resistance—Part 1: general rules, seismic actions and rules for buildings. EN 1998:2005

Dequaire X (2012) Passivhaus as a low-energy building standard: contribution to a typology. Energy Eff 5:377–391

DEUS 2010/31/EU. Direktiva o energetski učinkovitosti stavb [Online]. Available: http://eur-lex.europa.eu/LexUriServ/LexUriServ.do?uri=OJ:L:2010:153:0013:0035:SL:PDF. Accessed 12 July 2013

EU Commission (2019a) Comprehensive study of building energy renovation activities and the uptake of nearly zero-energy buildings in the EU [Online]. Available: https://ec.europa.eu/energy/studies_main/final_studies/comprehensive-study-building-energy-renovation-activities-and-uptake-nearly-zero-energy_sl. Accessed 13 May 2021

EU Commission (2019b) The European green deal. Belgium, Brussels

EU Commission (2021) The new European bauhaus [Online]. Belgium: Brussels. Available: https://europa.eu/new-european-bauhaus/index_en. [Accessed]

Hoxha E, Habert G, Lasvaux S, Chevalier J, le Roy R (2017) Influence of construction material uncertainties on residential building LCA reliability. J Clean Prod 144:33–47

Krainer A (2011) System. Module 1, concept of bioclimatic design (In Slovene). Lectures on the subject of buildings, energy, environment. University of Ljubljana, Faculty of Civil and Geodetic Engineering, Chair of Buildings and Constructional Complexes (KSKE)

Küçüktopcu E, Cemek B (2018) A study on environmental impact of insulation thickness of poultry building walls. Energy 150:583–590

Passivhaus Institut (2012) Certification criteria for residential passive house buildings [Online]. Darmstadt, Germany. Available: http://www.passiv.de/downloads/03_certfication_criteria_residential_en.pdf. Accessed 18 Aug 2013

UN (United Nations) (2015) UN sustainable development goals [Online]. Available: https://sustainabledevelopment.un.org/sdgs. Accessed 15 Jan 2021

Open Access This chapter is licensed under the terms of the Creative Commons Attribution 4.0 International License (http://creativecommons.org/licenses/by/4.0/), which permits use, sharing, adaptation, distribution and reproduction in any medium or format, as long as you give appropriate credit to the original author(s) and the source, provide a link to the Creative Commons license and indicate if changes were made.

The images or other third party material in this chapter are included in the chapter's Creative Commons license, unless indicated otherwise in a credit line to the material. If material is not included in the chapter's Creative Commons license and your intended use is not permitted by statutory regulation or exceeds the permitted use, you will need to obtain permission directly from the copyright holder.

Chapter 2
Design of the Thermal Insulation Envelope of Energy-Efficient Buildings

Among the crucial requirements to reduce energy consumption in buildings is a better thermal insulation envelope. It can be achieved by increasing the thickness of thermal insulation or improving the thermal transmittance of structural assemblies, eliminating all thermal bridges to ensure continuous thermal envelope. Thermal bridges are weak locations in the building envelope, where the thermal resistance (R) of a structural assembly is significantly lower than the thermal resistance of nearby locations in the building envelope. To recognise thermal bridges and eliminate their negative impacts, the theoretical knowledge of structural detailing and the calculation of heat transfer through the building envelope are required. The EN 10211 standard (CEN 2008) provides the definition of thermal bridges—locations in the building envelope where the otherwise uniform thermal resistance is significantly changed by:

- full or partial penetrations of the building envelope by materials with a different thermal conductivity;
- changes in the thickness of the structural assembly on the building envelope;
- differences between internal and external areas, such as occur at wall/floor/ceiling junctions.

Thermal bridges in the building envelope can be divided into various groups or subgroups based on their origin:

- geometrical (they occur at the locations of the building envelope, where the inner surface is smaller than the outer surface; this causes more intensive heat flow);
- structural or material (they occur due to changes in the thermal resistance of the building envelope; at various interruptions of the thermal envelope);
- convective (they occur in the building envelope where interruptions or a lack of air-tightness facilitate the flow of indoor moist air into the structural assembly); and
- their combinations.

Thermal bridges can also be divided by (sub)type:

- linear (thermal bridges with a constant cross-section along one of three orthogonal axes; in comparison with the total dimension of the building element (e.g. the width of a wall), they are narrow and stretch along the length which equals multiple thickness of the building envelope); and
- point (localised thermal bridges whose impact on total heat flow is captured in point thermal transmittance; most frequently, such thermal bridges are material related, for example for fixing façade elements with higher thermal conductivity).

Similar causes for the occurrence of thermal bridges and differences between their various types exceed the scope of this work. More information can be found in Déqué et al. (2001), Larbi (2005), Ge et al. (2013), Goulouti et al. (2014), Asdrubali et al. (2012), Martinez et al. (2017).

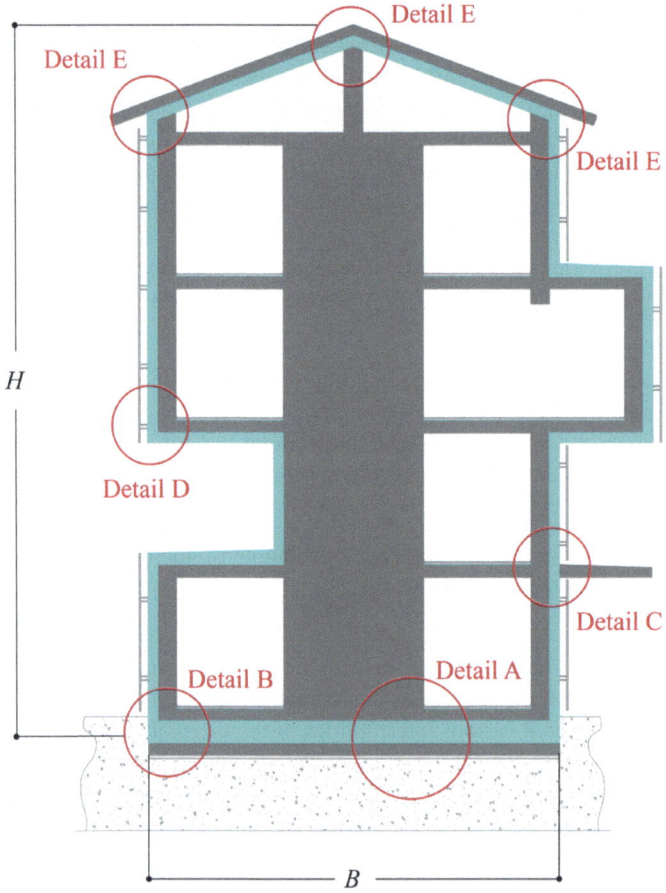

Fig. 2.1 Schematic representation of details in low-energy buildings, which are potentially critical from the aspect of earthquake resistance

Potentially seismically critical details in a building are (Fig. 2.1):

A. foundations on the thermal insulation layer;
B. special base insulation blocks for preventing thermal bridges in walls that are in contact with cold elements;
C. details for preventing thermal bridges in cantilever structures;
D. fixing façade elements;
E. fixing the roof structure and ensuring a stiff roof diaphragm.

The problem of thermal bridges must be solved in the design phase of a building or of a structural detail. Thorough design of building connections between structural assemblies, which are most critical locations where thermal bridges can occur, can completely prevent or reduce the influence of structural and geometrical thermal bridges or their combination (from the physical aspect, geometrical thermal bridges cannot be completely avoided). Potential locations of thermal bridges can be discovered in the conceptual design of a detail, whereby structural assemblies are treated as a combination of three basic layers—the load-bearing structure, thermal insulation, and the waterproofing layer. The basic principle that must be taken into account for the building envelope to function properly is that all basic layers must be uninterrupted and connected (Krainer 2002, 2011).

In the first step of the building connection solution, the load-bearing structure must be connected, then other layers in structural assemblies must be added (Fig. 2.2). In the second step, protective layers are provided to protect the structural assembly against external actions, such as temperature changes (warm, cold), rain, snow, groundwater, air humidity, etc. In the third step, it must be checked whether protective layers can perform their basic function (whether all layers are continuous and whether additional protective layers are required). At this point, we can determine whether a thermal bridge will occur, or the waterproofing layer will be interrupted in any of the details. In this way, we can avoid the use of such detail already in the conceptual design phase. In the fourth step, the load-bearing structure must be designed for all protective layers, and it must be checked whether a connection with the primary load-bearing structure is possible and if other protective layers are obstructed.

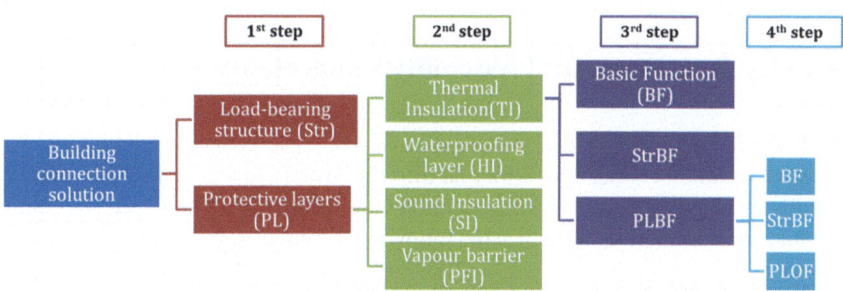

Fig. 2.2 Functional analysis of the building connection detail assembly. *Source* diagram summarised from Krainer and Kristl (2008), p. 9

To get a better idea, here is an example of thermal insulation whose basic function in the building envelope is to protect against heat transfer in the event of low temperatures outside, which is why it is treated as a protective layer. To enable thermal insulation to perform its basic function throughout the building lifetime, it must be suitably fixed (with various glues, anchors, screws, etc.), which is called the load-bearing structural system for thermal insulation in Fig. 2.2. This will prevent undesired behaviour of thermal insulation, such as collapse, separation, swelling, etc. In addition, other occurrences, such as condensation resulting from obstructed vapour diffusion, which increases thermal conductivity of thermal insulation, can prevent thermal insulation from performing its basic function. Therefore, a vapour barrier acting as a protective layer for thermal insulation must be installed in certain structural details (Fig. 2.2). All other protective layers (the waterproofing layer, sound insulation, etc.) must be analysed in a similar way as it was presented for thermal insulation.

We frequently use matrix presentations when studying conceptual solutions for structural assembly connections. Matrix presentations of conceptual solutions for structural assembly connections are found in various sources (DIN 2006; Krainer and Kristl 2008; Krainer 2002; Mittag et al. 2003), and are designed for planning and selecting structural details. The selected solutions from the matrix presentation of conceptual solutions for structural assembly connections have the scale of 1: 20 or even 1: 10, 1: 5 and 1: 1, which requires considerable accuracy and a more precise presentation of the detail. Selecting a detail concept whose solution foresees a thermal bridge does not mean that the detail is useless; the thermal bridge can still be reduced or prevented when addressed in more detail (the scale of 1: 20, 1: 10 …). Sections 2.1–2.3 include matrix presentations of conceptual solutions for structural assembly connections varied in their geometry (»L«, »T« and »+« connection types), whereby four structural assemblies were selected with three basic layers (the load-bearing structure, the waterproofing layer, and thermal insulation). The locations of recognised thermal bridges are marked with a red arrow (see Figs. 2.3, 2.4, and 2.5). Various solutions of such details as provided by experts for non-earthquake-prone areas are presented later in Chap. 3.

2.1 »L« Type Structural Assembly Connections

Figure 2.3 includes the matrix presentation of »L« type conceptual solutions for structural assembly connections with four different structural assemblies. The matrix presentation shows connections in the section view or floor plan. L-shaped structural assembly connections may appear at the following locations in the building envelope:

- between an outer wall and the foundation slab;
- between an outer wall and the flat roof;
- between an outer wall and an outer wall, etc.

2.1 »L« Type Structural Assembly Connections

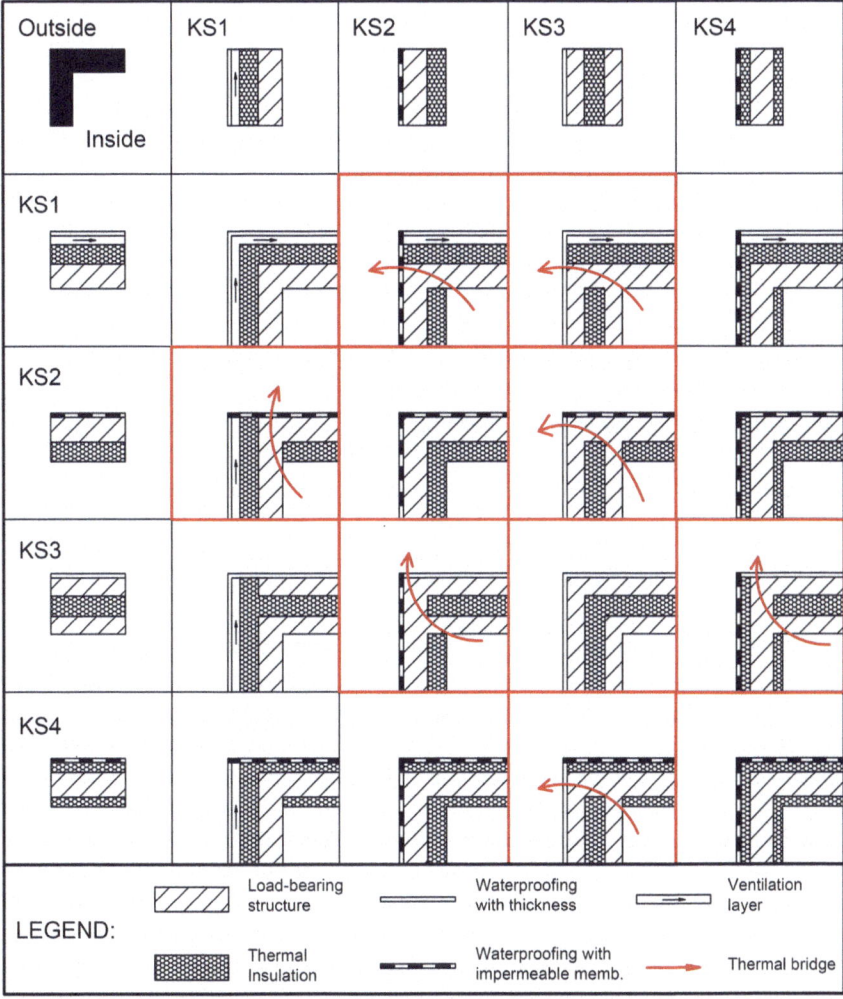

Fig. 2.3 Matrix presentation of »L« type conceptual solutions for structural assembly connections. *Source* Summarised from Krainer and Kristl (2008) and Krainer (2002, 2011)

Materials and the thickness of the load-bearing structure and protective layers are not defined in the conceptual design, but their position in the structural assembly is. The position of protective layers for the four selected structural assemblies is stated in Table 2.1 and schematically shown in Figs. 2.3, 2.4, and 2.5.

Based on the matrix presentation of »L« type connections, it can be established that details on the matrix main diagonal (for building connections between two identical structural assemblies) are always uncomplicated (without any interruptions in the protective layers of thermal insulation and the waterproofing layer). From the aspect of interrupted thermal insulation, seven problematic connections can be established,

Table 2.1 Conceptual design of the analysed structural assemblies

Assembly code	Thermal insulation position	Waterproofing	Ventilated layer
KS1	External side	With sufficient thickness	Yes
KS2	Internal side	With impermeable membrane	No
KS3	In the core of the load-bearing structure	With sufficient thickness	No
KS4	Covering the load-bearing structure from both sides	With impermeable membrane	No

which cannot be eliminated theoretically and require further consideration. It is characteristic of problematic connections that thermal insulation cannot be connected in the conceptual design without interrupting the load-bearing structure. The only option to prevent thermal bridges in such a building connection is to use a material that combines good thermal insulative and structural properties (e.g. sufficient load-bearing capacity). A careful consideration of the location where thermal insulation is interrupted is necessary in subsequent stages of designing such connection. Using appropriate materials in the building load-bearing structure (e.g. load-bearing thermal insulation elements or base insulation blocks), certain compromises must be made from the technical and structural aspect as presented below. If the solution with thermal insulation materials is not feasible, but we still want to use the detail in an energy-efficient building, we must calculate the temperature profile (energy flow) and determine the linear thermal transmittance (ψ) for the thermal bridge in the analysed connection detail. If the thermal bridge is significant, it must be taken into account in the calculation of energy consumption for the building. We must also verify how surface temperatures decrease in comparison with the indoor temperature. Thus, we can determine whether condensation and related mould will occur in this detail, and whether the detail is acceptable from the aspect of thermal comfort. Only when thermal analyses show that thermal bridge is acceptable and ensures thermal comfort and when the condensation is prevented, such detail could be used in energy-efficient buildings.

When analysing the »L« type connection between an outer wall and the foundation slab, solutions with the KS1 structural assembly for the foundation slab must be excluded, as the ventilated layer makes them useless. Additional attention must be paid to solutions with the KS4 structural assembly, since the thermal insulation in this case becomes the load-bearing layer in the foundation slab detail. At this point, thermal insulation with good thermal insulation properties, high stiffness and compressive strength must be reinstalled. More attention to the details of foundations on thermal insulation is paid in Sect. 3.1.

2.2 »T« Type Structural Assembly Connections

»T« type structural assembly connections can occur at the following locations in the building envelope:

- between an inner wall and the foundation slab;
- between a basement wall and the interstorey slab;
- between an inner wall and the roof;
- between an outer wall, the roof and the cantilever, etc.

To solve building connections at the theoretical or conceptual level, the part of the connection that borders the exterior and the part of the connection that is within the building envelope must be determined. By defining the relation between the outside and inside, we determine the part that must be protected with thermal insulation and the waterproofing layer. In addition, defining the heated part and the exterior affects the range and determination of thermal bridges. In Fig. 2.4, the wall structural assembly in the matrix in question is not treated as part of the building envelope, since the wall separates two heated rooms. Consequently, a structural assembly without thermal insulation could be used for such a wall. Nevertheless, the same structural assemblies as stated in Table 2.1 are shown. In this case, thermal insulation can act as sound insulation. This is frequently used in practice to separate two housing units in residential buildings and also two rooms heated at different temperatures.

The matrix presentation of »T« type conceptual solutions for structural assembly connections shows that five details are problematic from the aspect of the occurrence of thermal bridges. Similar to »L« type connections, thermal insulation is interrupted due to the primary requirement for the continuous vertical load-bearing structure. Regarding building connections between an inner wall and the foundation slab, base insulation blocks can be used in certain masonry structures, which are presented in more detail in Sect. 3.2. In »T« type connections which are connections between an inner wall and the roof, problematic details can be solved by inserting thermal insulation in certain sections to interrupt the load-bearing structure. In these cases, thermal bridges can still occur locally, but can be significantly reduced. In addition, more attention must be paid to the anchorage of the roof structure and its connection to outer or inner walls for full functioning in the event of the horizontal seismic loads as described in more detail in Sect. 3.4. We want to reiterate building connections with the KS1 structural assembly for the foundation slab, which are not useful due to the ventilated layer. Particular attention must also be paid to KS4 if this is a structural assembly with thermal insulation under the foundation slab.

2.3 »+« Type Structural Assembly Connections

The most complex matrix solution for »+« type connections which can occur in the building envelope at the following locations: between an outer wall, the interstorey

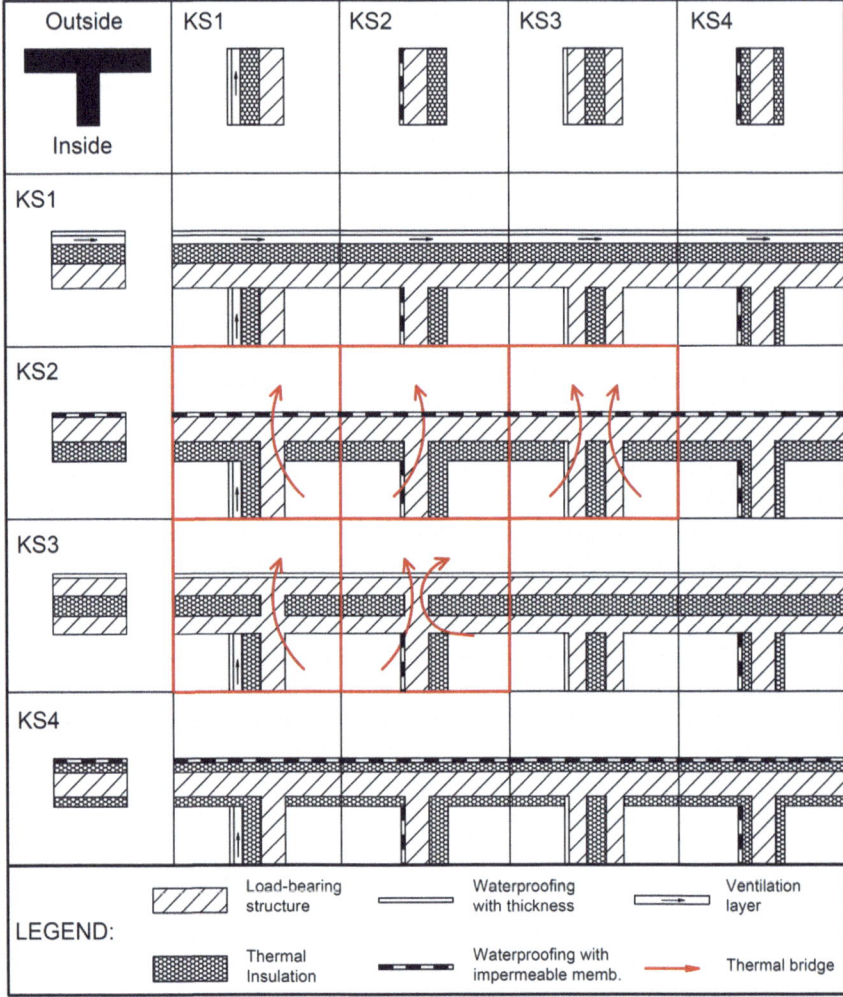

Fig. 2.4 Matrix presentation of »T« type conceptual solutions for structural assembly connections. *Source* Summarised from Krainer and Kristl (2008) and Krainer (2002, 2011)

slab and the balcony cantilever or the loggia; between a basement wall, the inter-storey slab and an inner wall, etc. Matrix solutions in Fig. 2.5 showed nine building connection details where a thermal bridge was present at the conceptual level. If any of such problematic connections was to be used (e.g. for a balcony cantilever), load-bearing thermal insulation elements can be used or the fact that the thermal bridge does not significantly affect energy consumption in the building must be proven. These solutions are presented in more detail in Sect. 3.3 and in the Appendix.

2.3 »+« Type Structural Assembly Connections

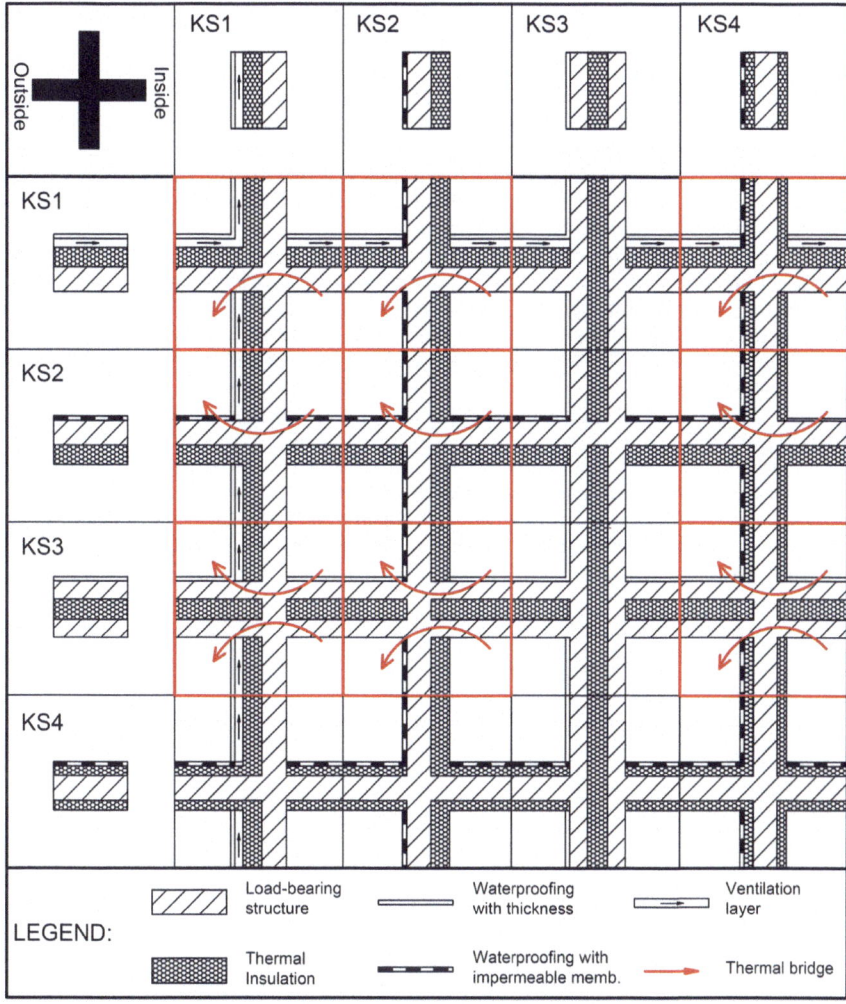

Fig. 2.5 Matrix presentation of »+« type conceptual solutions for structural assembly connections. *Source* Summarised from Krainer and Kristl (2008) and Krainer (2002, 2011)

References

Asdrubali F, Baldinelli G, Bianchi F (2012) A quantitative methodology to evaluate thermal bridges in buildings. Appl Energy 97:365–373

CEN (2008) Thermal bridges in building construction—heat flows and surface temperatures—detailed calculations. EN ISO 10211:2008. CEN

Déqué F, Ollivier F, Roux J (2001) Effect of 2D modelling of thermal bridges on the energy performance of buildings: numerical application on the Matisse apartment. Energy Build 33:583–587

DIN (2006) Thermal insulation and energy economy in buildings—thermal bridges—examples for planning and performance. DIN 4108 Beiblatt 2:2006-03. Beuth, Berlin

Ge H, McClung VR, Zhang S (2013) Impact of balcony thermal bridges on the overall thermal performance of multi-unit residential buildings: a case study. Energy Build 60:163–173

Goulouti K, de Castro J, Vassilopoulos AP, Keller T (2014) Thermal performance evaluation of fiber-reinforced polymer thermal breaks for balcony connections. Energy Build 70:365–371

Krainer A (2002) Roofs. Module 1, structural connections 4 (In Slovene). Lectures on the subject of Buildings. University of Ljubljana, Faculty of Civil and Geodetic Engineering, Chair of Buildings and Constructional Complexes (KSKE)

Krainer A (2011) System. Module 1, concept of bioclimatic design (In Slovene). Lectures on the subject of Buildings, energy, environment. University of Ljubljana, Faculty of Civil and Geodetic Engineering, Chair of Buildings and Constructional Complexes (KSKE)

Krainer A, Kristl Ž (2008) Connections (In Slovene). Lectures on the subject of Buildings. University of Ljubljana, Faculty of Civil and Geodetic Engineering, Chair of Buildings and Constructional Complexes (KSKE)

Larbi AB (2005) Statistical modelling of heat transfer for thermal bridges of buildings. Energy Build 37:945–951

Martinez RG, Riverola A, Chemisana D (2017) Disaggregation process for dynamic multidimensional heat flux in building simulation. Energy Build 148:298–310

Mittag M, Trbojević R, Jovanović V, Jovanović J, Stanišić M (2003) Građevinske konstrukcije: Priručnik za graditelja o konstruktivnim sistemima, građevinskim elementima i načinima gradnje, Građevinska knjiga

Open Access This chapter is licensed under the terms of the Creative Commons Attribution 4.0 International License (http://creativecommons.org/licenses/by/4.0/), which permits use, sharing, adaptation, distribution and reproduction in any medium or format, as long as you give appropriate credit to the original author(s) and the source, provide a link to the Creative Commons license and indicate if changes were made.

The images or other third party material in this chapter are included in the chapter's Creative Commons license, unless indicated otherwise in a credit line to the material. If material is not included in the chapter's Creative Commons license and your intended use is not permitted by statutory regulation or exceeds the permitted use, you will need to obtain permission directly from the copyright holder.

Chapter 3
Structural Details in Energy-Efficient Buildings

When designing energy-efficient buildings, building designers encounter various details and challenges, such as the building orientation, the position and size of openings, the architectural room layout, a good building thermal envelope, and numerous others (Lukić et al. 2019; Premrov et al. 2015; Passivhaus Institut 2012; Manzano-Agugliaro et al. 2015). This chapter highlights structural details intended to prevent thermal bridges and heat losses through the building envelope.

3.1 Foundations on the Thermal Insulation Layer

The use of thermal insulation under the foundation slab first became widespread when passive houses were developed 20–30 years ago. Only the definition of the requirements of the PH standard fostered the determination of the minimum requirements for the thermal transmittance of the slab on ground, which cannot be achieved without thermal insulation. In addition to the requirement on minimum thermal transmittance, one of the main requirements of the PH standard is the prevention of thermal bridges (this Chapter). As shown in the L and »T« type matrix solutions, the only solution without a thermal bridge is to use thermal insulation under the foundation slab or alternatively base insulation blocks in masonry structures. This has triggered new developments of foundation slabs on thermal insulation layers.

3.1.1 Possible Solutions for Foundations on Thermal Insulation

In current practice, there are two systems to thermally insulate the foundation slab or shallow foundations under the building (Fig. 3.1). One way is to replace the soil and the drainage layer in the foundation base with insulation material (Fig. 3.1a),

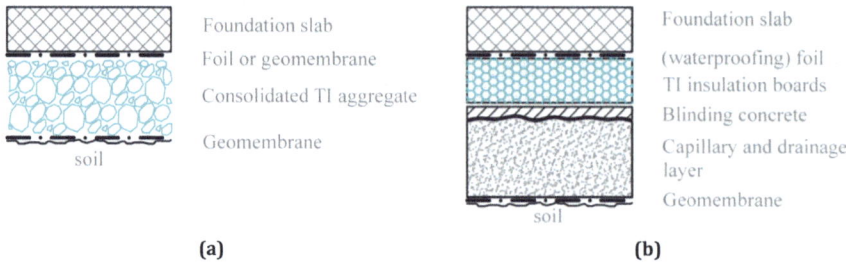

Fig. 3.1 Comparison of TI foundation systems: **a** construction with aggregate and **b** construction with TI boards. *Source* Summarised from Kilar et al. (2014a)

while the other is to install one or more layers of thermal insulation boards under the foundation slab (Fig. 3.1b). This section includes a review of materials that can be used for thermal insulation boards and for thermal insulation aggregate. The characteristics and technical properties of all available materials are presented, and the advantages and disadvantages of both systems are pointed out.

Foundations on thermal insulation

The review and properties of aggregates were summarised from various manufacturers and data in their catalogues, and relevant scientific literature (Lyons 2014; GLAPOR 2020; Weber 2015; Hess Perlite 2015; Dupre 2015; Zegowitz 2010; Ozguven and Gunduz 2012; Demirboğa and Gül 2003; Janetti et al. 2015). Aggregate fill materials installed under the foundation slab must have high compressive strength, as they carry the load of the entire building and are part of the load-bearing foundation base. At the same time, aggregates in thermal insulation must be highly stable and have the anti-capillary effect. The most important properties of aggregates include a low thermal conductivity and a good compression of the aggregate, which improve thermal insulation properties. A considerable disadvantage of insulation aggregates in comparison with thermal insulation boards is higher thermal transmittance in the event of high groundwater reaching above the aggregate benchmark. Therefore, most manufacturers recommend aggregate to be 30 cm higher than the highest groundwater point.

Table 3.1 shows the properties of aggregate insulation fill materials most frequently used in construction, which include cellular glass, expanded clay, vermiculite, and perlite. Based on the properties stated in the Table 3.1, we can argue that not all aggregate insulation fill materials used in construction are suitable for insulation under the foundation slab. Vermiculite and perlite were included in the review of aggregate insulation fill materials on the assumption that their load-bearing capacity is high, as they are used an additional aggregate in lightweight insulation concrete (Sengul et al. 2011; Kan and Demirboğa 2009). Nevertheless, vermiculite and perlite aggregate grains showed minimum compressive strength, which accounts for poorer properties of insulation concrete, which reach a lower compressive strength than conventional concretes. Since vermiculite and perlite do not demonstrate sufficient compressive strength, we believe that they cannot be used as insulation under the

3.1 Foundations on the Thermal Insulation Layer

Table 3.1 Typical properties of aggregate insulation fill materials used for building applications

Material property*	Cellular glass	Expanded clay	Expanded vermiculite	Expanded perlite
Density ρ (kg/m^3)	100–150	220–300	70–120	40–170
Thermal conductivity λ (W/(m K))	0.060–0.080	0.065–0.120	0.035–0.120	0.025–0.050
Granulation (mm)	10–60	10–30	0–15	0–6
Granulate compressive strength (kPa)	400–1600	700–2500	10–50	10–50
Shear angle (°)	40–50	35–40	–	–
Cohesion c (kPa)	0	0	–	–
Energy for production** (kW h)	85	73–168	11–24	–
Relative material costs**	5.3–5.9	5.9–12.8	1.8–2.6	–

*There is no uniform standard to determine the properties of aggregate insulation fill materials, meaning data can vary slightly, as they were obtained with various methods
**Energy for production and relative material costs are standardised with equivalent thermal transmittance U = 0.4 W/(m^2 K) (AURE 2003; IBO 2008). Relative costs are standardised by the price of EPS (Table 3.2)
Source Summarised from: Lyons (2014), GLAPOR (2020), Weber (2015), Hess Perlite (2015), Dupre (2015), Zegowitz (2010)

foundation slab. These two materials have small grains that are significantly smaller than the grains of the other two analysed materials (cellular glass and expanded clay). Vermiculite can be used for interstorey slab insulation (it is also soundproof), but as such does not act as a load-bearing layer but merely as a filler between primary beams (Dupre 2015). Perlite can be used in a similar way and also under the slab on ground. Perlite-filled bags can be used for insulation under the floor slab, but are not the load-bearing layer of the entire building but only of the floor slab (Hess Perlite 2015).

According to the properties of compressive strength, cellular glass and expanded clay can be used under the foundation slab. However, there is a notable difference between their thermal conductivity, i.e. at the same density/compressive strength, the thermal conductivity of expanded clay is significantly higher than of cellular glass. In addition, significantly more energy is required to produce high-tensile and insulating aggregate from expanded clay than from cellular glass (Table 3.1), resulting in a higher price of the insulating aggregate from expanded clay. A disadvantage of expanded clay is that it requires a thicker layer of insulation to achieve the same thermal transmittance of the slab on ground structural assembly than cellular glass and other aggregate fill materials. Considering a higher price and a negative environmental impact (IBO 2008), we can say that the most appropriate material for insulation under the foundation slab in current practice is cellular glass. In addition to the properties from Table 3.1, the positive properties of cellular glass are water

Table 3.2 Properties of thermal insulation boards used in floor applications (screed, floorboard, flat roof, and foundation slab)

Material property	Cellular glass	Extruded polystyrene (XPS)	Expanded polystyrene (EPS)	Polyurethane (PUR/PIR)	Mineral wool
Density ρ (kg/m^3)	100–165	25–35	15–30	30–100	40–200
Thermal conductivity λ (W/(m K))	0.040–0.065	0.030–0.040	0.031–0.043	0.020–0.035	0.03–0.045
Water repellency W_{lp} (% of volume)	<0.2	<0.3	<1	<1.6	<3
Compressive strength σ_{10} (kPa)	400–1600	100–1000	30–500	25–800	10–90
Strength at compressive creep σ_{cc} (kPa)	100–700	20–300	10–150	5–250	2–30
Elastic modulus E (MPa)	100–500	15–40	5–25	2–25	0.3–2
Shear strength τ (kPa)	80–400	100–200	10–300	100–450	5–50
Shear modulus G (MPa)	<4	3–8	1.5–9	1–5	0.3–1.5
Energy for production* (kW h)	85	43–89	39–95	47–64	9–90
Relative material costs*	5.3–5.9	3–3.5	1–1.2	<3	1–1.5

*Energy for production and relative material costs are standardised with equivalent thermal transmittance U = 0.4 W/(m^2 K) (IBO 2008). Relative costs are standardised by the price of EPS
Source Summarised from: JUBHome (2016), Fibran (2020), Foamglas (2020), Rockwool (2020), Elfoam (2020)

resistance (individual aggregate grains), high durability, and anti-capillary action. A literature review (Foamglas 2015) showed that cellular glass insulation had already been used for insulation under large buildings, meaning that this is not just insulation for single-family houses generally characterised by low effects on the ground.

Cellular glass aggregate is produced from waste or recycled glass with mineral additives with heat treatment at over 900 °C, during which powder from the basic

3.1 Foundations on the Thermal Insulation Layer

production compound is shaped into 5–8 cm thick boards. As they cool to room temperature, they disintegrate into the final product, i.e. cellular glass aggregate, due to residual stress from the heat treatment process. During this process, foam-like aggregate grains with a closed cell structure are formed. In addition to insulation under the foundation slab, cellular glass aggregate is also useful, due to its high compressive strength and low density, in other building applications: for perimeter insulation in industrial or commercial buildings; for load-bearing and vehicular areas in multi-storey car parks; for the perimeter insulation of floors or buildings; for the insulation of arcades, arches or bridge piles; and for antifreeze materials in surface underground ducts. The installation of insulative aggregate as used for cellular glass aggregate is shown in Fig. 3.2 (GLAPOR 2020).

Expanded clay is less widely used as thermal insulation under the foundation slab. Expanded clay (Arioz et al. 2008; Laterlite 2007) has so far been most frequently used to reduce the self-weight of other load-bearing materials. The lightweight grains of expanded clay are mixed with soil or concrete to avoid the problem of considerable self-weight. For example, expanded clay under the foundation slab is used in areas with low load-bearing capacity of the ground. In this way, the same load was preserved as before the building construction, protecting the ground from collapsing. In a similar way, self-weight can be reduced by using expanded clay in lightweight concrete. Using expanded clay in lightweight aggregates is not the same as using expanded clay in insulation. The latter must be done with a suitable geomechanical protection as shown in Fig. 3.2. Literature on expanded clay insulation aggregate is scarce, which leads us to conclude that it is not frequently used. In addition, the manufacturers of

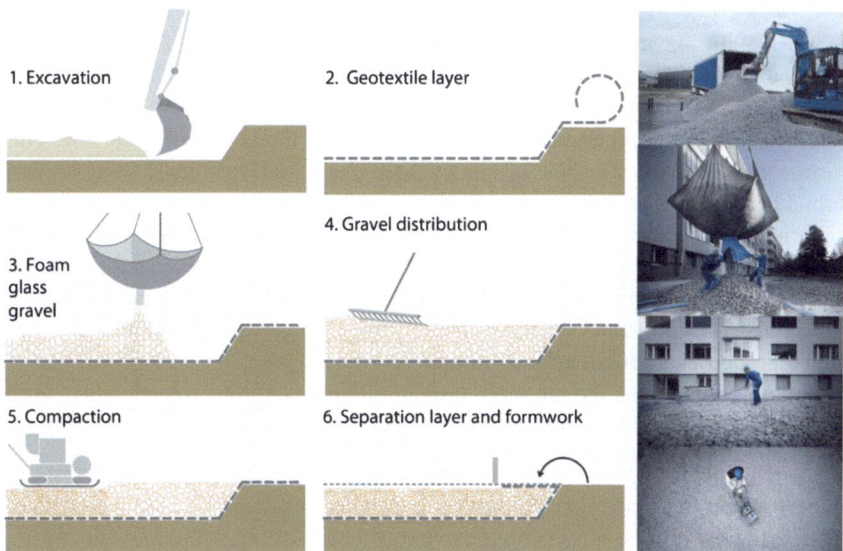

Fig. 3.2 Delivery and installation of cellular glass gravel insulation. *Source* Summarised from GLAPOR (2020)

such a material (Weber 2015) limit the use of the material for thermal insulation only for small buildings. The review of details with expanded clay insulation aggregates showed that most insulation aggregates are not placed under the vertical load-bearing structure but under the floor slab, which means a significantly lower compressive load.

Relevant scientific literature does not explore in detail the impact of insulation aggregates under the foundation slab on the structural safety of buildings. From the scientific research point of view, the topic relates to geotechnics (shallow foundations) and also building structures. Out of the four presented aggregate insulation fill materials (cellular glass, expanded clay, vermiculite, and perlite), only cellular glass corresponds to the strict requirements of application under the foundation slab. Therefore, additional research options are the development of new competitive materials and improvements in the production processes of aggregates with insulation properties. Experts in geomechanics and structural engineers should cooperate to explore whether insulation aggregates affect the load-bearing capacity of the ground or whether such aggregates can compromise the ground properties. Another concern regarding insulation aggregates under the foundation slab is connected with the travelling of seismic waves and consequences of the seismic response of the building above. As stated before, insulation aggregates are also used for multi-storey buildings in less earthquake-prone areas. If the technology of insulation aggregate under the foundation slab for buildings in earthquake-prone areas was to be transferred, the highlighted problem should be investigated and these research questions should be suitably responded. Some research questions were recently addressed by Banović et al. (2020), Brandis et al. (2021).

Foundations on thermal insulation boards

Like insulation aggregates, thermal insulation boards used under the foundation slab are exposed to various external actions. In addition to the discussed properties of insulation aggregates (low thermal conductivity, high compressive strength, and durability), important properties of thermal insulation boards include shear strength, deformability (shear, compressive), compressive creep, and water resistance. Certain thermal insulation boards preserve virtually the same thermal conductivity and other properties even at constant moisture. Therefore, they can be used even if they are in constant contact with the soil and at high groundwater levels. However, if the material properties change significantly due to moisture, we must place the waterproofing layer in a structural assembly under thermal insulation, ensuring the material has optimal properties throughout its lifetime.

Shallow foundations on thermal insulation boards differ considerably from insulation aggregates. The installation procedure can be summarised as follows (Fig. 3.3):

- construction pit excavation and inserting installations under the foundation slab;
- preparing the ground (preloading, gravelling and levelling with blind concrete);
- installing thermal insulation (it may be installed in one or more layers) and the waterproofing layer (the designer determines its position in the structural assembly depending on the water resistance of thermal insulation);

3.1 Foundations on the Thermal Insulation Layer

Fig. 3.3 Example of extruded polystyrene (XPS) insulation boards used under the foundation slab. *Source* Summarised from Fibran (2020)

- installation of concrete formwork, placing of steel reinforcement and placing/curing the concrete of the foundation slab.

In Table 3.2, the properties of thermal insulation materials in boards used for horizontal structural assemblies (e.g. interstorey slab, foundation slab and roof) are shown. The review and properties of thermal insulation materials were summarised from various manufacturers and data in their catalogues, and relevant scientific literature (JUBHome 2016; Fibran 2020; Foamglas 2020; Rockwool 2020; Ramsteiner et al. 2001; Bunge and Merkel 2011; Méar et al. 2007; Diascorn et al. 2015; Gnip et al. 2011, 2010; Elfoam 2020; Papadopoulos 2005). Strength and deformation-related properties in Table 3.2 were obtained by static monotonic tests. Therefore, the obtained values can be used for designing of structures in non-earthquake-prone areas and for static forces, while dynamic and cyclic experiments must be carried out for more in-depth analyses of the seismic response. In comparison with insulation aggregates, thermal insulation boards have been used for the thermal insulation of the building envelope for a long time. In the EU, there are harmonised standards determining essential requirements for insulation boards for the purpose of use in construction and to ensure insulation materials conformity (CEN 2013c, d, e, f, g). In addition to thermal conductivity, the main requirement for thermal insulation boards in horizontal structural assemblies (insulation of walk-on roofs, insulation under the foundation slab and (or) screed) is sufficient compressive strength. The compressive strength values quoted in Table 3.2 were obtained using the EN 826 (CEN 2013a) and EN 1606 (CEN 2013b) standards, which distinguish two compressive stresses, i.e. maximum compressive strength σ_m (if the diagram has no explicit maximum strength, σ_m is determined at the deformation of ten percent, hence the mark σ_{10}) and the highest permissible compressive stress for permanent loads σ_{cc} (determined for the 50-year lifetime with a two-percent deformation).

The values in Table 3.2 show that rigid mineral wool boards are not appropriate for thermal insulation under the foundation slab, as their compressive strength is too low and deformability too high (low values of the elastic and shear modulus). They can also be excluded due to a high water absorption rate, which is an important characteristic for the durability of a structural assembly in contact with the soil. Another material that can be ruled out as a material to be used under the foundation slab is polyurethane. Its strength and deformation-related properties, and price are comparable to those of EPS and XPS, and it even shows lower thermal conductivity. Nevertheless, PUR/PIR is not used under the foundation slab, since its production entails various environmental risks (propellants that are harmful to the ozone layer were used to produce rigid foams from PUR/PIR until recently). Section 4.2.6 contains more information on environmental parameters and the life cycle assessment. Such a method of exclusion leaves three materials that can be used as thermal insulation under the foundation slab, and are most frequently encountered on the market (XPS, EPS, and cellular glass).

Figure 3.4 shows the microstructure of the three discussed materials, which can be used for additional comparison of these materials. EPS and XPS boards have the same basic material (polystyrene), but differ in the way they are produced (extrusion and expansion). Since they expand in a closed space, polystyrene grains are polyhedral in shape and glued together due to high temperatures during steaming (JUBHome 2016; Papadopoulos 2005). By contrast, the structure of XPS boards (Fig. 3.4a) is more homogeneous, as they are made by extrusion (Modic 2009). This largely affects water absorption, which is higher in EPS boards. For this reason, XPS boards have an advantage over EPS boards in structural assemblies in constant contact with moisture (inverted roof, basement wall, foundation slab, etc.), as they preserve low thermal conductivity regardless of the presence of moisture. Research into EPS boards used as so-called geofoams for reducing the self-weight of the soil (Athanasopoulos et al. 1999; Vejelis et al. 2008; Gnip et al. 2007; Maleki and Ahmadi 2011; Forcellini 2020; Yoshihara et al. 2018; Yoshihara and Maruta 2020) has provided answers regarding their durability and preservation of good mechanical properties when in constant contact with the soil and moisture (strength and deformability). On the other hand, the foundation slab structural assembly must be suitably waterproofed if EPS boards are used as thermal insulation under it. This prevents high thermal conductivity of EPS on account of water content, and ensures optimal operation throughout the building lifetime. In comparison with EPS and XPS, the microstructure of cellular glass is similar to that of XPS, as it is characterised by a closed cell structure and good water repellency (Fig. 3.4).

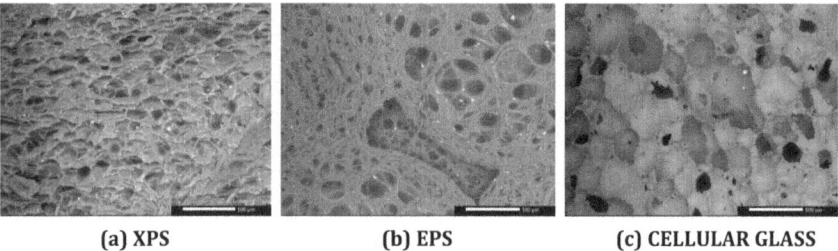

(a) XPS   (b) EPS   (c) CELLULAR GLASS

Fig. 3.4 Microstructure of thermal insulation materials. *Source* Slovenian National Building and Civil Engineering Institute (author: Andreja Pondelak)

3.1 Foundations on the Thermal Insulation Layer

The comparison of strength and deformation-related properties of the three discussed materials (XPS, EPS, and cellular glass) revealed the poorest properties of EPS, which is, nevertheless, frequently used, as it is significantly cheaper than XPS and cellular glass (Table 3.2). In addition, the production process of EPS boards facilitates the production of other useful shapes with moulds. Therefore, it can also be used instead of formwork for concrete work for the reinforced concrete foundation slab, which can speed up the construction process. Manufacturers (JUBHome 2016) permit the construction of houses founded on EPS boards for single-storey and multi-storey lightweight structures (e.g. prefabricated timber buildings, autoclaved aerated concrete) and up to two storeys for solid construction (reinforced concrete and brick masonry buildings). XPS and cellular glass, which exhibit greater mechanical resistance, are recommended for taller and more exposed buildings. Cellular glass has higher compressive strength and elastic modulus than other insulation materials, but also the highest thermal conductivity. Its weakness is also its high price, which is why it is only used when other materials are not appropriate (e.g. for the insulation of multi-storey complex buildings).

In general, designers of thermal insulation under the foundation slab prefer foundations on thermal insulation boards to insulation aggregates on account of the following advantages:

- less material is required for the same thermal insulation effect, since insulation aggregates have higher thermal conductivity than thermal insulation boards;
- they are more appropriate when groundwater is high, as thermal insulation boards can be protected with a waterproofing layer, preserving low thermal conductivity throughout the building lifetime;
- their installation is cheaper and simpler—mainly on the account of EPS and XPS, which are the cheapest materials from the list and can be produced only as boards.

Preventing thermal bridges and impact on surface temperatures

This section shows improvements in the environmental and energy-efficiency parameters depending on the thickness of thermal insulation under the foundation slab for the selected structural »L« type connection detail between an outer wall and the foundation slab. The temperature range and heat flow in the 2D section of the analysed details had to be determined to assess the influence of thermal insulation under the foundation slab on the environmental and energy-efficiency parameters of the details. The method used to determine the influence of a thermal bridge and the environmental and energy-efficiency parameters of the detail is presented in Sects. 4.2.1–4.2.4. Details with different thickness of thermal insulation under the foundation (0–45 cm) were compared with the temperature range analysis. Boundary conditions (location and climatic conditions) used in detail calculation are prescribed for Central Europe and its continental climate (indoor design temperature of 20 °C, outside temperature of -10 °C, 60-percent relative humidity, etc.). The assumed type of soil is gravel, since this is the most common layer under the foundation slab of a building used for local reinforcement of the ground and a well-drained layer.

In the reference case, insulation is not installed under the foundation slab as required for energy-efficient or passive buildings. The only thermal insulation is located under screed, and also acts as sound insulation against impact sound. The reference case without thermal insulation is summarised from detail *DT*-02 presented and assessed from the technical and structural, and environmental and energy-efficiency aspects (see the appendix below). More details on the composition of structural details and materials used are included in the description of detail *DT*-02. The weak spots of uninsulated detail *DT*-02, which indicate the course of a thermal bridge (locations marked with red and orange signify a higher heat flow rate), can be deducted from the diagram of heat transfer in this detail. At these locations, heat transfers most rapidly to the ground. The temperature drop on the detail surface is the highest at the location of the highest heat flow rate. As expected, this happens on the corner or connection between the wall structural assembly and the foundation slab due to a combination of a geometrical and structural thermal bridge at this location.

We further modified the detail *DT*-02 by adding thermal insulation under the foundation slab of various thicknesses (Fig. 3.5), and determined the temperature range and the course of heat flow for all the analysed details. Figures 3.6 and 3.7 show the lowest surface temperature ($\theta_{si,min.}$) and linear thermal transmittance of the thermal bridge (ψ) depending on the thickness of thermal insulation reached in all the details. It can be deducted from Fig. 3.6 how the minimum surface temperature of a detail increases with increasing thickness of thermal insulation under the foundation slab ($\theta_{si,min.}$). As expected, the most significant difference is in the initial thickness (up to 10 cm), where the slope of the curve is the steepest. With increasing

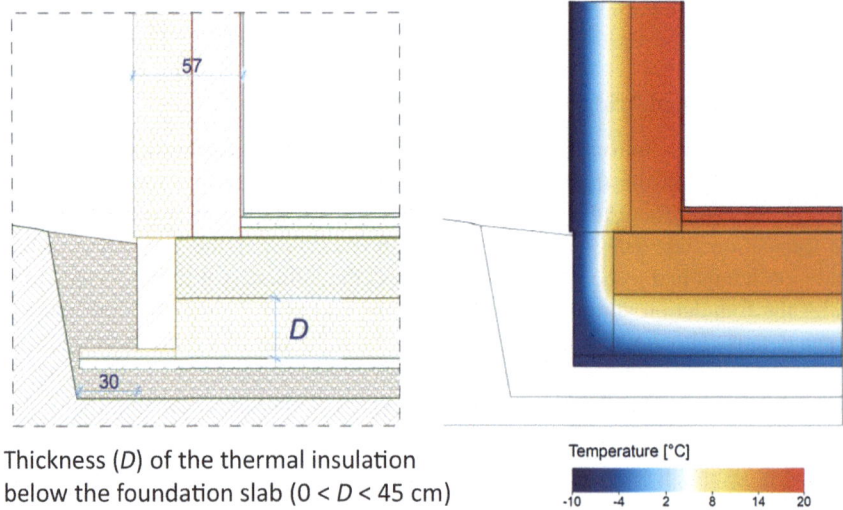

Fig. 3.5 Analysed building connection between an outer wall and the foundation slab according to thermal insulation thickness and the results of heat transfer simulation for the thermally insulated foundation slab detail

3.1 Foundations on the Thermal Insulation Layer

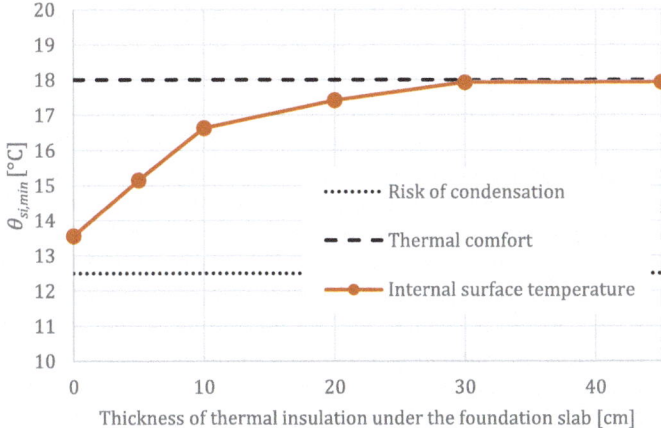

Fig. 3.6 Influence of thermal insulation under the foundation slab on the increase of the internal surface temperatures of the building »L« type connection between an outer wall and the foundation slab

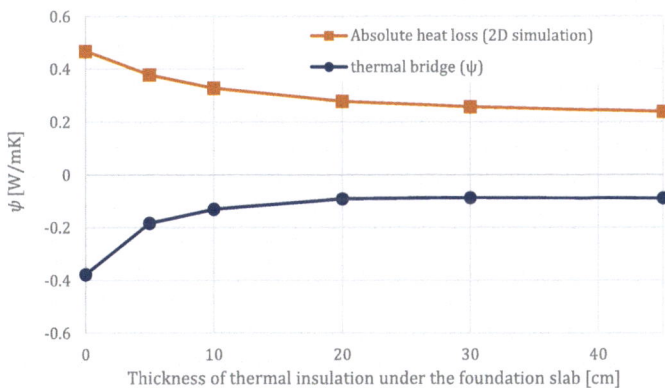

Fig. 3.7 Influence of thermal insulation under the foundation slab on the linear thermal transmittance of thermal bridges (calculated for external dimensions)

thickness of thermal insulation, the increase in surface temperatures is not as significant. The temperature difference between details with thermal insulation thickness between 20 and 45 cm is merely 0.5 °C. Figure 3.6 also shows guidance values for condensation and thermal comfort. Condensation and related mould greatly depend on boundary conditions (in the discussed study, indoor design temperature is 20 °C, outside temperature is −10 °C, and relative humidity is 60%). Temperature range simulations for the analysed boundary conditions did not reveal any risks of condensation even if a structural assembly has no thermal insulation under the foundation slab.

Table 3.3 Thermal transmittance of the insulated foundation slab according to thermal insulation thickness

Thermal insulation (cm)	None	5	10	20	30	45
U (W/(m^2 K))	0.52	0.29	0.21	0.13	0.10	0.07

If we increase the thickness of thermal insulation under the foundation slab, in addition to a better temperature range, heat losses of the entire detail are reduced (see the orange curve in Fig. 3.7). Heat losses are highest up to the thermal insulation thickness of 10 cm. Therefore, thicker thermal insulation has no significant effect and installing over 30 cm of insulation is questionable due to minimal effects on the thermal response. The influence of the heat losses of a detail can be divided into losses resulting from thermal transmittance of structural assemblies ($\sum_i A_i U_i$), linear thermal bridges ($\sum_k l_k \psi_k$), and from point thermal bridges ($\sum_j \chi_j$). Most heat losses in the analysed detail are losses resulting from the thermal transmittance of structural assemblies (Table 3.3). In all cases, thermal insulation on the external side of the building envelope is continuous (apart from structural assemblies without thermal insulation), preventing structural (material) thermal bridges. Others are a result of the linear geometrical thermal bridge marked with a blue curve in Fig. 3.7.

In better structural assemblies with greater thermal insulation thickness, relative losses resulting from a linear geometrical thermal bridge are higher. Such a result is expected and supports the fact that the elimination of thermal bridges is even more important in energy-efficient buildings with a well-insulated thermal envelope. Since all the analysed cases include the same detail type, in which only the thickness of thermal insulation changes, heat flow at the location of a linear geometrical thermal bridge intensifies if the thickness of thermal insulation under the foundation slab increases. A similar phenomenon may be encountered in the case of the energy renovation of the building envelope (e.g. adding thermal insulation on the inside). In practice, only the thickness of the thermal insulation on the building envelope is frequently increased during energy renovation without eliminating structural and other thermal bridges. Therefore, heat flow at locations of thermal bridges relatively increases (Marincioni et al. 2015; Krause et al. 2020).

3.1.2 Specifics of Designing Thermally Insulated Foundation Slabs in Non-earthquake-Prone Areas

This section focuses on the structural safety of buildings founded on thermal insulation, which are mainly exposed to vertical static actions and wind, whereby seismic action is not relevant, since the buildings are located in non-earthquake-prone areas. In the context of designing building structures for static actions and wind, thermal insulation under the foundation slab produces certain specifics in comparison with the thermally uninsulated foundation slab. We must be aware that thermal insulation

3.1 Foundations on the Thermal Insulation Layer

under the foundation slab (as insulation aggregates or insulation boards) becomes part of the load-bearing foundation base for the whole building, which must be taken into account by structural engineers to prove structural safety in all foreseen load situations throughout the building lifetime. Instructions for designing thermal insulation under the foundation slab and related specifics have not been provided yet within the Eurocode standards of structural design, as this is a relatively new technology. Whenever structural design standards (e.g. Eurocode) do not directly address a problem, the requirements for mechanical resistance and building stability can be met by taking into account the principles and sensible use of these standards, technical guidelines, and other technical documents if it is possible to use them to meet the requirements on at least an equivalent level. The basic standards to determine actions on structures (e.g. actions determined in accordance with the Eurocode 1 standards (CEN 2004b): imposed loads for buildings, snow and wind loads) and load safety factors (e.g. according to the Eurocode 0 standard (CEN 2004a)) must still be considered.

It is thus sensible to rely on the provisions from regulations to determine the rules for designing the load-bearing layer of thermal insulation. This chapter provides an approach for designing the load-bearing layer of thermal insulation, which stems from the provisions of the Eurocode standards. In this case, the rules from Section 3.5 (fill, dewatering, ground improvement and reinforcement) of Eurocode 7: Geotechnical design (CEN 2005b) are most relevant to thermal insulation aggregates. Particular attention must be paid to a potentially uneven settlements of the foundation base from insulation aggregates and compression procedures must be supervised. More information on designing thermal insulation in the form of boards is stated below. The following must be additionally checked in buildings with a layer of thermal insulation boards installed under the foundation slab:

- exceeding the maximum (compressive and shear) resistance of thermal insulation boards under the foundation slab in the ultimate limit state calculation;
- long-term creep of thermal insulation boards; and
- design of the reinforced concrete (RC) foundation slab on elastic foundation.

When verifying static equilibrium and resistance, the following basic equation from Eurocode 0 should be taken into account:

$$E_d \leq R_d, \qquad (3.1)$$

whereby E_d is the design value of the effects of actions, such as internal forces and moments, and R_d is the design value of the resistance.

Controlling thermal insulation boards in the ultimate limit state

Design stress in thermal insulation boards in non-earthquake-prone areas depends on vertical static actions (self-weight and permanent loads, imposed loads, snow loads) and horizontal wind actions. When controlling compressive stress due to highest short-term load (ultimate limit state—ULS), it must be ensured that the maximum compressive resistance of thermal insulation boards is not exceeded. It is also recommendable to prevent their irreversible deformation. To determine vertical stress, it must be ensured that no tensile stress occurs under the foundation slab, which would result from eccentricity due to horizontal loads (e.g. wind). In this case, this section of the foundation slab must be excluded from the contact surface in line with the design practice. The compressive resistance of thermal insulation boards is verified with the following equation:

$$\sigma_d \leq \gamma_i \, \sigma_{10}/\gamma_m, \qquad (3.2)$$

whereby σ_d is compressive stress due to a combination of vertical and horizontal non-seismic actions in ULS, σ_{10} is the compressive strength of thermal insulation boards (Table 3.2), γ_m is the material safety factor, and γ_i is an additional safety factor to prevent irreversible compressive deformations.

Determining the material safety factor of load-bearing thermal insulation boards is not regulated. Material safety factors largely depend on the basic raw material of insulation boards and the process of their manufacture. Therefore, it is generally difficult to determine a common safety factor applicable to all materials. The values of the material elastic modulus also significantly differ, which means that safety factors for preventing irreversible deformations must be determined for each analysed material individually. Certain instruction for the use of safety factors were prepared by manufacturer JUBHome (2016) for EPS, relying on the publication EPS White Book (EUMEPES 2014) to harmonise the structural use of EPS with the Eurocode standards. In this publication, $\gamma_m = 1.08$ is proposed to control compressive strengths in ULS, which can be adopted for material safety factors. In addition, the value of the safety factor of EPS boards was provided to prevent irreversible deformations $\gamma_i = 0.40$. Safety factors when thermal insulation boards from other materials shown in Table 3.2 are used should be determined in a similar way.

In addition to the compressive stress in ULS, also shear stresses of the compressive part of thermal insulation boards (τ_d) should be verified, which may occur due to horizontal forces on the building (wind actions). The control of shear stresses is written as the control of compressive stresses (Eq. 3.2):

$$\tau_d \leq \tau_{\min.}/\gamma_m, \qquad (3.3)$$

whereby τ_d is design shear stress in ULS in the compressed section, $\tau_{\min.}$ is the ultimate shear resistance of the slab on ground structural assembly, and γ_m is the material safety factor.

3.1 Foundations on the Thermal Insulation Layer

The ultimate shear resistance of a structural assembly may refer to the shear strength of a board if the structural assembly only contains one layer of thermal insulation boards or if sliding at the contact between individual layers in the structural assembly is prevented. On the other hand, $\tau_{min.}$ can mean maximum shear stress caused by sliding between individual layers in the structural assembly (contact between thermal insulation boards if they are installed in several layers or at the contact between thermal insulation boards and other layers, e.g. blinding concrete, waterproofing layer). Sliding would occur if the static friction factor on the surface between individual layers in the structural assembly was exceeded. For $\tau_{min.}$, the lower of both values is used. Practice has shown that, the ultimate shear resistance is in most cases higher than design shear stress, which is a result of wind actions in non-earthquake-prone areas. In addition, Eq. (3.3) does not take into account additional safety due to the passive resistance of the soil in the case of an underground structure, which is why the control of the ultimate shear resistance is not relevant.

Long-term creep of thermal insulation boards

In addition to maximum static actions addressed by Eurocode 0 as part of ultimate limit state control, creep control in the long-term compressive load is also important when designing thermal insulation boards for vertical static loads. Material creep means an increase in its deformations with constant stress caused by long-term loads. For thermal insulation (TI) materials in the European harmonised standards demand that a deformation due to creep does not exceed two percent in the building lifetime (50 years). Compressive stress that leads to a two percent deformation in 50 years is called strength at compressive creep (σ_{cc}). Strength σ_{cc} is determined with long-term tests in which deformation at constant stress is measured and values are extrapolated for presumed life-time periods (e.g. 50 years). To control the long-term creep, the long-term elastic modulus (E_{cc}), which is determined with tests according to compressive stresses that lead to creep, and measured deformations must be known. The Eq. (3.2) for the ultimate limit state can be transformed into a form applicable to creep control:

$$\sigma_d \leq \sigma_{cc}/\gamma_m, \tag{3.4}$$

whereby σ_d is design compressive stress resulting from a combination of long-term external actions.

Design compressive stress resulting from long-term loads (σ_d) in the sense of Eurocode 0 is determined with safety and combination factors for the so-called quasi-permanent combination of actions determined for the serviceability limit state (SLS). Control in the Eq. (3.4) has an adverse effect due to long-term creep, such as uneven settlements of the foundation slab. Practice shows that a great deal of compressive stress is a result of self-weight and dead load (i.e. long-term loads leading to creep), which for most TI materials signifies the relevance of static control due to the long-term creep. The strength σ_{cc} of the materials in Table 3.2 is on average three times lower than compressive strength σ_{10}.

Modelling reinforced concrete foundation slab on elastic foundations

Structurally, a thermally insulated reinforced concrete (RC) foundation slab is a floor slab on elastic foundations, to which a layer of thermal insulation boards gives additional flexibility, which must be taken into account during analysis and design. Different available calculation methods can be used to calculate bending loads and reinforcement in RC slabs. In practice, linear (beam) or 2D (shell) finite elements on springs, which constitute vertical flexibility resulting from the deformation of thermal insulation boards and soil flexibility, are most frequently used. The calculation model is the basis for determining internal forces in the foundation slab, which is dimensioned according to the rules from the standard in the design of concrete structures Eurocode 2 (CEN 2005a). Sufficient stiffness and load-bearing capacity of the foundation slab must be ensured to distribute the loads resulting from the vertical load-bearing structure on the layer of thermal insulation and soil as evenly as possible.

A common modulus of subgrade reaction for determining the characteristics of vertical springs of the numerical model is calculated as the reciprocal value of the sum of reciprocal values from the modulus of soil reaction k_z and the contribution of thermal insulation k_{TI} (determining stiffness for consecutive springs) (Merkel 2004; JUBHome 2016). The vertical modulus of soil reaction k_z is usually provided by a geomechanics specialist or is assessed with the expected settlement of the soil (without thermal insulation), which is dependent on the size of the foundation slab and also on the level of vertical loads. Attention must be paid to differences between absolute and relative or local settlements under the foundation slab. The magnitude of absolute settlements under a building is not essential. However, even minor relative (differential) settlements influence the design of a foundation slab and its reinforcement. A common modulus of subgrade reaction (k_v) stands at:

$$k_v = 1 / \left(\frac{1}{k_z} + \frac{1}{k_{TI}} \right) \quad (\text{kN/m}^3). \tag{3.5}$$

The contribution of thermal insulation in the calculation of the modulus of subgrade reaction (k_{TI}) is obtained as follows:

$$k_{TI} = \frac{\sigma}{\Delta u_{TI}} = \frac{\sigma}{\varepsilon \, d_{TI}} = \frac{E_{TI}}{d_{TI}} \quad [\text{kN/m}^3], \tag{3.6}$$

whereby σ is compressive stress in thermal insulation, Δu_{TI} is the settlement of thermal insulation, ε is compressive deformation of thermal insulation, d_{TI} is the thickness of thermal insulation, and E_{TI} is the elastic modulus of thermal insulation. The Eqs. (3.5) and (3.6) apply on the assumption of a linear elastic response (not applicable to inelastic response).

To calculate the elastic modulus of thermal insulation (E_{TI}), E or E_{cc} can be used, depending on which structural design control is relevant. For example, the long-term modulus E_{cc} is relevant to the design of a slab under a combination of long-term actions.

3.1 Foundations on the Thermal Insulation Layer

Limitations of founding buildings on thermal insulation in non-earthquake-prone areas

Previous sections show that in non-earthquake-prone areas designers must focus on the control of compressive stress in thermal insulation (TI) boards and the design of RC foundation slab located on a flexible TI layer. In most cases of compressive stress control, the control of long-term stresses leading to the creep of thermal insulation turns out to be relevant. Such a control is relevant particularly on account of significant deformability of materials used in thermal insulation boards (Table 3.2) with constant stress (low elastic modulus E_{cc}). This can be particularly hazardous at locations of compressive stress concentrations (where walls, columns are fixed), where larger settlements than in the rest of the foundation slab can occur in the event of more flexible foundation slabs. To prevent uneven settlements brought on by creep, the Eq. (3.4) must be taken into account at all locations of stress concentration. In addition to structural limitations, the following actions must be considered when determining thermal insulation under the foundation slab: water resistance of the material (preserving good properties in the presence of water, protection with a waterproofing layer); resistance to freezing and melting; adverse environmental impacts (sourcing of raw materials, processing, recycling); sensitivity to hydrocarbon fuels and other solvents (if there is a risk of such exposure, protection must be provided with a suitable lining or foils); durability or ageing of the material; risk of microorganisms; etc.

Based on the structural limitations described above, the greatest challenge for structural engineers in non-earthquake-prone areas is to limit the compressive stress in the thermal insulation layer. Increases in the compressive stress depends on the building structural system, the number of storeys, the load-bearing structure material (lightweight and solid construction), the intended use of the building (scale of the imposed load), location (size of snow and wind actions), underground section (with a basement), floor plan dimensions, etc. Limitations to the number of storeys (n) in non-earthquake prone areas could be estimated with a simple calculation in the context of the Eq. (3.4) to limit long-term stress and by taking into account material limitations stated in Table 3.2 on the assumption that the stress under the foundation slab is constant. To determine design compressive stress (σ_d), the equations from the SLS for the quasi-permanent combination of actions are used and a building of medium weight and a solid RC load-bearing structure is selected. 15 kPa is used for the vertical stress of a storey ($\sigma_{n,d}$), which is only a rough estimate, since the weight of a storey may vary significantly and must be calculated in more detail in a real project (Kilar et al. 2013). XPS 700 with strength at compressive creep $\sigma_{cc} = 250$ kPa and material safety factor $\gamma_i = 1.25$ are presumed for the thermal insulation material (DIBT 2013). Based on these assumptions, compressive creep control (Eqs. (3.4) and (3.7)) shows that XPS could theoretically withstand a 13-storey building:

$$n < \frac{\sigma_{n,d}}{\sigma_{cc}} \gamma_i = \frac{15\,\text{kPa}}{250\,\text{kPa}} \, 1.25 = 13.3 \quad n < 13. \tag{3.7}$$

Nevertheless, the effect of wind actions and other eccentricities requires that designers have to be more conservative if such high-rise buildings are to be built on a layer of thermal insulation boards. Other reasons are also additional stress concentrations and specifics that cannot be covered by a simplified calculation with the Eq. (3.4) or (3.7). The other reason is connected with energy—annual energy savings in high-rise buildings with the insulated foundation slab are not as high, which in practice means a longer payback period for the insulated layer of the building.

3.1.3 Influence of the Flexible Layer of Thermal Insulation on the Seismic Response of a Building

The construction of energy-efficient buildings with thermal insulation under the foundation slab has been present in earthquake-prone areas for some time. In addition to the specifics and limitations of earthquake-prone areas, changes in seismic response may be expected resulting from the installation of thermal insulation under the foundation slab. This section highlights potential specifics of buildings with thermal insulation under the foundation slab in earthquake-prone areas. It includes a brief review of scientific literature depicting known state of the art and current findings about the seismic response of buildings founded on a flexible base (i.e. flexible soil, seismic isolation, thermal insulation).

The seismic response of buildings founded on thermal insulation is somewhat similar to the soil-structure interaction (SSI). In both cases, excessive settlement of the foundation slab, sliding and rocking of the building (Fig. 3.8), the separation of

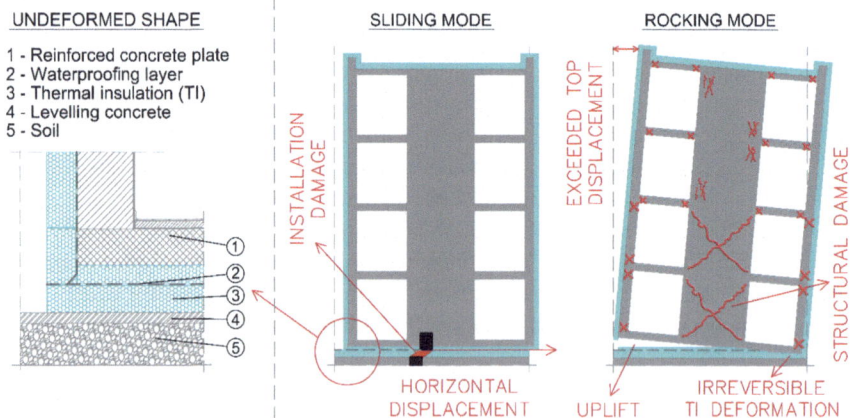

Fig. 3.8 Assumed seismic response of energy-efficient buildings founded on thermal insulation

3.1 Foundations on the Thermal Insulation Layer

the foundation slab from the ground (exclusion of the tensile zone), and the seismic energy dissipation due to hysteretic action of the ground could occur. The literature review of important findings about the soil-structure interaction is provided in Raychowdhury and Hutchinson (2009), Lou et al. (2011), Raychowdhury (2011), Kourkoulis et al. (2012b), Gazetas et al. (2013), Dhadse et al. (2020), Anand and Kumar (2018), Cavalieri et al. (2020). In recent years, numerous studies have shown that the effects of the SSI can be crucial to the seismic response of a structure and must be considered for more exposed buildings (founded on very flexible soil, highrise and slender buildings, etc.) and in more complex seismic analyses. The latter is also prescribed in Eurocode 8. The effects of the SSI on certain types of soil and structures has proven to be favourable and could reduce seismic requirements in comparison with structures on rigid ground (e.g. Nakhaei and Ali Ghannad (2008), Raychowdhury (2011), Gelagoti et al. (2012a), Jarernprasert et al. (2013), Banović et al. (2020)). Other studies (e.g. Mylonakis and Gazetas (2000), Dutta et al. (2004), Mahmoud et al. (2012)) have shown that the effects of the SSI must be taken into account when analysing certain low-rise buildings, as the SSI could lead to higher internal forces and damage to the upper structure.

The effects of the SSI may be taken into account with linear and nonlinear springs for the so-called Winkler foundation or a half-space (Chopra and Yim 1985; Dutta et al. 2004; Allotey and El Naggar 2008; Roy and Chandra Dutta 2010) or with complex three-dimensional (3D) models that consider both the soil and the upper structure (Gazetas et al. 2007; Gelagoti et al. 2012b; Kourkoulis et al. 2012c; Anastasopoulos and Kontoroupi 2014). More accurate results may be obtained with 3D analyses. However, they frequently turn out to be challenging in terms of calculation and time-consuming, which is why they are mostly replaced by simpler Winkler spring models that still foster sufficient accuracy. A modified Winker spring model can also be used for the foundation slab on thermal insulation, where the properties of the flexible soil must be replaced with the properties of thermal insulation boards. The topics of the SSI and thermal insulation under the foundation slab are similar, however, numerous studies on the effects of the SSI in the event of an earthquake do not provide complete and sufficient answers regarding the seismic response and limitations of buildings founded on thermal insulation. Considerable differences lie in the changed characteristics used for the modelling of the flexible layer (thermal insulation characteristics instead of soil) and a rather thin layer (up to 30 cm) of thermal insulation in comparison with the half-space taken into account in soil modelling. The structural assembly of the load-bearing foundation slab, which is placed on two or more layers of load-bearing thermal insulation boards, is also changed.

In certain respects, the flexible layer of thermal insulation under the foundation slab can be compared to similar seismic isolation systems (see, for example, Naeim and Kelly (1999), Skinner et al. (1993)). In the case of seismic isolators of elastomeric bearings (rubber bearings—RB, lead rubber bearings—LRB, etc.), their main characteristic is the extended fundamental period of a building (see, for example, Christopoulos et al. (2006), Kilar and Koren (2009)), producing lower seismic forces (Fig. 3.9b). Extended fundamental period can also be expected if

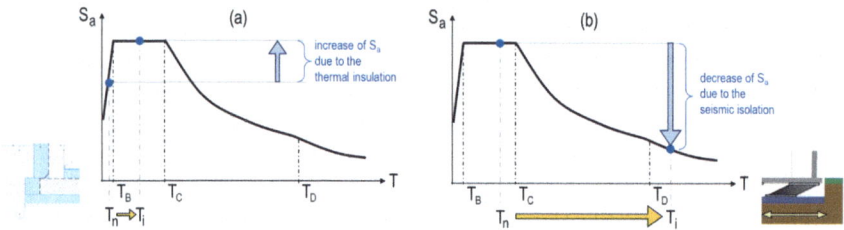

Fig. 3.9 Roof accelerations of the stiff superstructure on **a** thermal insulation (XPS); **b** conventional seismic isolation (rubber bearings) under the foundation slab. *Source* Kilar et al. (2013)

thermal insulation is installed under the foundation slab (Fig. 3.9a). The difference between both methods is that the fundamental period of seismic isolation is extended particularly on account of the shear deformability of elastomeric bearings, and in thermal insulation particularly on account of vertical deformability of thermal insulation boards.

Certain similarities in the seismic response could also be observed in buildings founded on friction pendulum base isolation stems (FPS) (Zayas et al. 1990; Panchal and Jangid 2009; Lu et al. 2013; Tsai et al. 2014; Becker and Mahin 2012; Fadi and Constantinou 2010; Chung et al. 2015; Timsina and Calvi 2021; Auad and Almazán 2021). Similar to elastomeric bearings, the fundamental period of such base isolated building is also extended. The functioning principle of friction pendulum systems in the event of strong earthquake lies in the sliding mechanism, which prevents stronger horizontal seismic forces on the upper structure, protecting it from major damage. The sliding mechanism must be controlled so that the building returns to its original position after an earthquake, which is achieved in certain types of sliding isolation with a curved surface of isolators. In this case the building could balance itself only by its self-weight. Also possible are other types of sliding isolation with added devices to return a building to its equilibrium position. The sliding mechanism could also occur in the case of certain foundation slab structural assemblies on several layers of thermal insulation due to a low friction coefficient of the contact surface between layers. The difference in comparison with sliding isolation is that the sliding surface in thermal insulation is not bent or additionally protected from sliding, which could lead to uncontrolled horizontal shifts of foundations. In addition to the conventional sliding seismic isolation, other methods applying the sliding mechanism to reduce seismic forces (e.g. using a smooth synthetic base in foundations, protecting masonry walls by enabling sliding at the contact with the waterproofing layer, etc.) can be found in scientific literature (Mojsilović et al. 2010; Yegian and Kadakal 2004; Nanda et al. 2012).

Certain assumptions on potential negative effects of thermal insulation we confirmed already in 2013 (see for example Kilar et al. (2013)). Inserting the thermal insulation layer under the foundation slab extends the fundamental period of a structure which might be moved into the resonance part of the design response spectrum

3.1 Foundations on the Thermal Insulation Layer

where the seismic forces are the largest (Fig. 3.9a). We have found that the fundamental period is not extended as much as with conventional seismic isolation, in which various devices (elastomeric bearings, friction pendulum systems) intentionally extend the fundamental period of a building by twice or more, significantly moving it away from the unfavourable resonance area (Fig. 3.9b). It was established that the extension of the fundamental period due to the installation of thermal insulation under the foundation slab can be particularly critical in energy-efficient buildings with a short fundamental period (approx. 0.1 s), since greater seismic forces can be expected in such buildings (Fig. 3.9a).

Preliminary studies provided certain evaluations regarding materials selected for thermal insulation boards. The assumption was that stronger seismic excitation can result in increased compressive stresses on the edges of the foundation slab (see Fig. 3.8). Analyses have shown that certain achieved compressive stresses exceed the nominal compressive strengths of thermal insulation, which is currently used in practice as insulation under the foundation slab (Fibran 2020). In view of the study assumption, more slender and heavyweight buildings (e.g. concrete) have proven to be the most problematic, in which the compressive strength of XPS-with the nominal compressive strength of 400 kPa was exceeded at the height of three storeys. According to the criterion on the compressive strength of XPS, lightweight buildings with a more favourable ratio between floor plan dimensions (e.g. timber buildings with floor plan dimensions ratio of 1:2–1:3) facilitate up to six storeys. Regarding the effect of the material (mass) of the load-bearing structure, construction with timber load-bearing elements generally allows for at least one storey more than construction with a concrete structure (for comparable floor plans of buildings). The control of maximum shear stresses and maximum horizontal displacements of XPS has not proven to be critical.

Dynamic and cyclical experimental tests of thermal insulation boards made of XPS and various configurations of structural assemblies of the foundation slab were carried out in Kilar et al. (2014b). A detailed parametric seismic study is shown in Azinović et al. (2014b; c), which is an upgrade of preliminary research and takes into account the nonlinear model of the upper structure and nonlinear thermal insulation layer. The potential use of the thermally insulated foundation slab detail for reducing seismic forces with controlled sliding is presented in Azinović et al. (2015b, 2016), Kilar et al. (2016).

3.2 Base Insulation Blocks for Preventing Thermal Bridges in Walls in Contact with Cold Elements

3.2.1 Types of Base Insulation Blocks

The term 'base insulation block' denotes the type of masonry units combining good thermal insulation properties (with a low thermal conductivity λ) and high compressive strength. They are used exclusively in masonry structures in combination with clay, concrete or autoclaved aerated concrete masonry blocks. From the environmental and energy-efficiency aspect, they are a solution for thermal bridges at the contact between a masonry wall and cold elements. These locations, particularly at the contact between an outer or inner wall and the unheated or cold basement (T and + connection types), between the strip foundation and an outer or inner wall (T and + connection types), and between an outer wall and a thermally uninsulated foundation slab (L connection type). Such thermal bridges can be most effectively prevented with thermal separation or the installation of base insulation blocks, depending on the location of the separation. Later, we focus on base insulation blocks used instead of conventional masonry units in the first line of a masonry wall (Fig. 3.10).

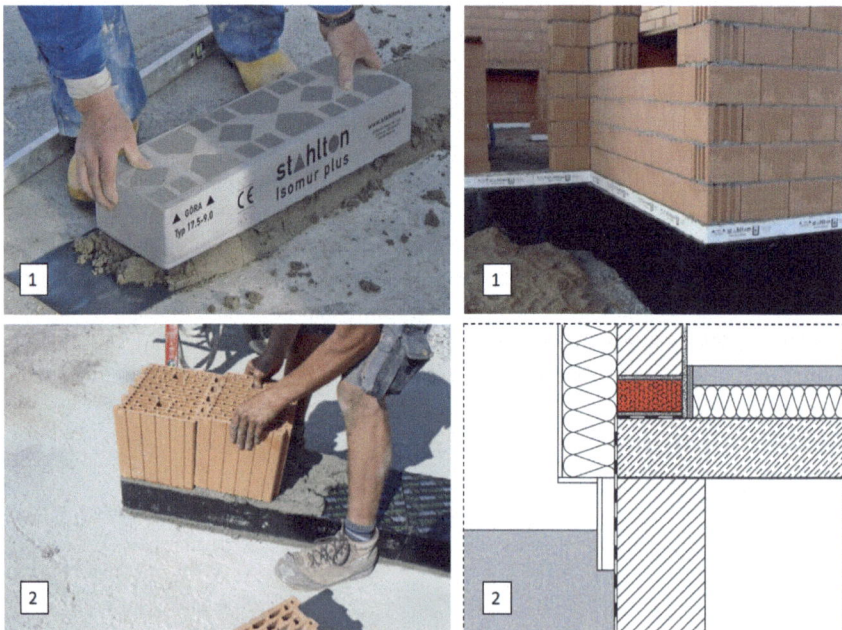

Fig. 3.10 Examples of a so-called insulation base for preventing thermal bridges at the junction of an outer/inner wall with cold surfaces (strip foundations, uninsulated foundation slab, unheated basement). *Source* https://www.stahlton-bauteile.ch/ (1) and https://www.foamglas.com/de-de (2)

3.2 Base Insulation Blocks for Preventing Thermal Bridges …

Considering the current practice, base insulation blocks can be roughly divided in two types (homogeneous and composite base insulation blocks). As the name suggests, homogeneous base insulation blocks are made of one material making up a homogeneous and uniform structure of a brick. Such base blocks include base blocks made of cellular glass, XPS, and autoclaved aerated concrete or foam concrete. The second type of base insulation blocks is made of two or more different materials. Certain materials have a load-bearing function and others have the thermal insulation function. Possibilities for composite base blocks are various and include base blocks which are a composite of brick and stone wool, brick and perlite, and various combinations of load-bearing (nano)concrete and thermally insulating XPS/EPS. The main difference between homogeneous and composite base insulation blocks is that the compressive strength of composite base blocks is usually higher. In relation to base insulation blocks, it is also important to distinguish thermal conductivity in the horizontal and vertical (i.e. load-bearing) direction of a masonry unit. Characteristic of homogeneous base insulation blocks is that their thermal conductivity is the same in both directions, while the thermal conductivity of composite base insulation blocks is higher in the vertical (i.e. load-bearing) direction. At the contact between the load-bearing wall and an uninsulated or cold slab, thermal bridges extend through the wall in the vertical direction. Therefore, it is crucial to reduce thermal conductivity in this direction, preventing intensive heat flow and improving the temperature range of the detail.

3.2.2 Influence of Base Insulation Blocks on Better Environmental and Energy-Efficiency Parameters

Figure 3.11 shows horizontal thermal conductivity of the most common base insulation blocks in relation to their density (Summarised from catalogues Marmox (2015), Schöck (2015), Wienerberger (2016), Ytong (2021), Fibran (2020)). Most manufacturers only state the lowest (i.e. horizontal) value of thermal conductivity, which is inadequate data from the aspect of preventing thermal bridges at the contact between the load-bearing wall and an uninsulated or cold slab, since in this case vertical thermal conductivity is crucial. Regarding composite base insulation blocks, for which all data were available, it was found that vertical thermal conductivity is on average higher than its horizontal counterpart by approximately one third. The values can be adapted to the exponential function showing how thermal conductivity increases with higher material density. Such a principle results from the fact that the best conductivity is characteristic of materials with high air content captured in a closed cell structure (e.g. EPS, XPS, etc.).

In a similar fashion, Fig. 3.14 shows normalised compressive strength of base blocks, to which the same principle applies to a certain extent (increasing density also increases compressive strength). Nevertheless, certain materials stand out, as they are characterised by high compressive strength at a low density and a low thermal

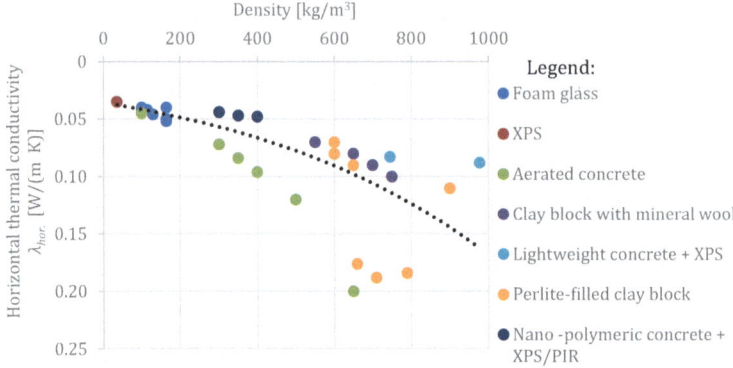

Fig. 3.11 Horizontal thermal conductivity of the most common base insulation blocks in relation to their density

conductivity. The basic requirement in the development of base insulation blocks is the best possible compromise between good properties of compressive strength and thermal conductivity.

The influence of base insulation blocks on environmental and energy-efficiency parameters are shown for the connection detail between an outer masonry wall and the reinforced concrete (RC) interstorey slab above the unheated basement (Sect. 5.3). The detail above the unheated basement is challenging, as thermal bridges must be prevented in two directions: towards the outdoor air and towards the unheated basement. For the building envelope to be continuous, using base insulation blocks at the location where the vertical masonry structure is fixed is almost the only possible option. Other options to improve the environmental and energy-efficiency parameters of the building connection detail between an outer wall and the unheated basement are not as efficient as base insulation blocks, as they merely reduce thermal bridges instead of eliminating them (e.g. by extending thermal insulation to the unheated basement, with additional thermal insulation on the internal side of the unheated basement, etc.).

To compare the environmental and energy-efficiency parameters of the analysed structural detail, we changed the value of the thermal conductivity of the insulation base block, and observed how the temperature range and the linear thermal transmittance change. Section 5.3 includes more on the composition of the structural assemblies and boundary conditions of the detail. Figure 3.12 shows the lowest temperature on the inner surface of the detail ($\theta_{si,\min.}$), while Fig. 3.13 shows the linear thermal transmittance coefficient for heat flow towards the unheated basement (ψ_2). The detail's temperature range can significantly improve with the insertion of an insulation base block. The first step included the analysis of the detail for which the thermal conductivity of the base block was assumed to be 2.5 W/(m K), which corresponds to the characteristic of the RC structure. It may be stated for such

3.2 Base Insulation Blocks for Preventing Thermal Bridges …

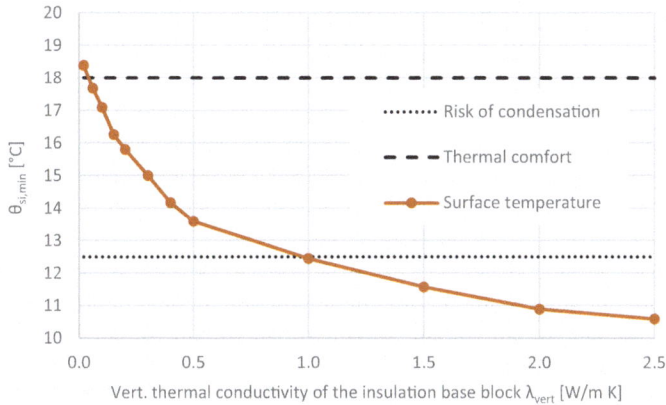

Fig. 3.12 Influence of base insulation blocks on the internal surface temperatures of the building connection between an outer wall and the unheated basement

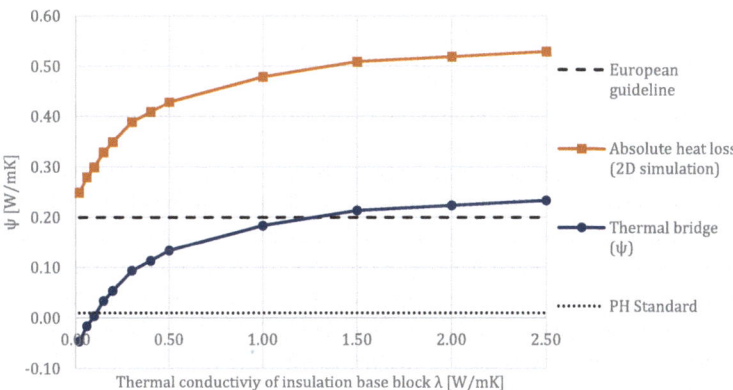

Fig. 3.13 Influence of the thermal conductivity of base insulation blocks on the linear thermal transmittance of thermal bridges (calculated for external dimensions) in relation to heat flow to the unheated basement

a detail that there is a strong possibility that condensation and related mould will occur on the inner surface, since the achieved temperatures are very low (approx. 10 °C) despite well insulated structural assemblies, which comply with the PH standard ($U < 0.15 \text{ W}/(\text{m}^2 \text{ K})$) and were used in the analysed detail (Sect. 5.3). In other models, the thermal conductivity of the base blocks was reduced to the limit that can still be achieved with advanced thermal insulation, moving towards the thermal conductivity of the insulation on other parts of the envelope (in the vertical and horizontal structural assemblies).

As shown in Fig. 3.12, temperatures over 18 °C could be reached on the inner surface of the detail with a very low thermal conductivity of the insulation base block in the best case scenario. Table 3.4 shows that only few base blocks (e.g. cellular

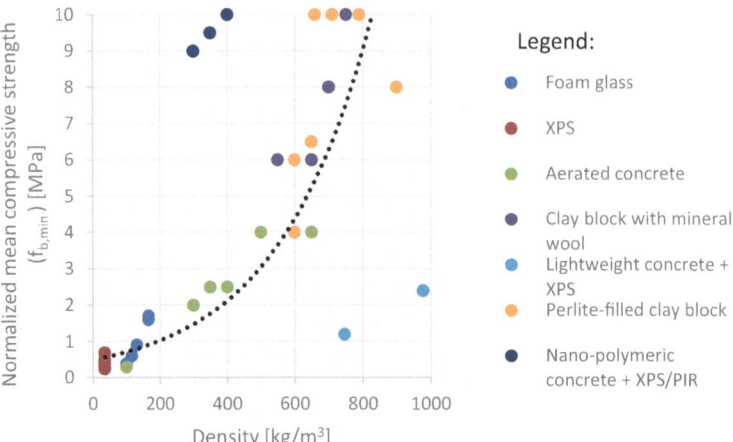

Fig. 3.14 Normalised compressive strength of the most common base insulation blocks in relation to their density

glass base blocks) reach a low thermal conductivity. Therefore, slightly lower surface temperatures (15–17 °C) can be expected in reality, which is still acceptable for use in modern energy-efficient buildings.

Figure 3.13 shows heat losses towards the unheated basement. It shows the value of heat losses for the whole detail (losses resulting from heat transfer through the structural assemblies) and the value of heat losses resulting from a thermal bridge towards the unheated basement. Unlike in the study of the influence of thermal insulation under the foundation slab, in which we changed the structural assembly (the thickness of thermal insulation), the study of the influence of the thermal conductivity of the insulation base block showed that the shape of the curve on relative losses (only due to a thermal bridge) is the same as the shape of the curve on heat losses of the whole detail. Such a result was to be expected, as thermal transmittance and dimensions of structural assemblies do not change, making the value of losses resulting from heat flow through structural assemblies constant.

The graph in Fig. 3.13 shows guidance values for the ultimate linear thermal transmittance of thermal bridges. The first limit value is summarised from proposals for European technical guidelines (e.g. MOP (2010)), which recommend avoiding thermal bridges with factor $\psi > 0.20\,\text{W}/(\text{m K})$. The second value stems from the PH standard, which states that thermal bridges with factor $\psi < 0.01\,\text{W}/(\text{m K})$ do not have to be taken into account in the calculation of energy use. The analysis of heat transfer showed that only cases with the thermal conductivity of base insulation blocks cca. $\lambda_{vert} < 0.10\,\text{W}/\text{mK}$ reach a very low value of linear thermal transmittance ($\psi < 0.01\,\text{W}/(\text{m K})$). In reality, such base insulation blocks with a very low vertical thermal conductivity and good compressive strength are difficult to produce. Therefore, higher values of heat losses are to be expected (approx. up to $0.10\,\text{W}/(\text{m K})$).

Table 3.4 Characteristics of homogeneous and composite base insulation blocks

Type of base blocks	Material	Trade name	Density ρ (kg/m³)	Hor. ther. cond. $\lambda_{hor.}$ (W/(m K))	Vert. ther. cond. $\lambda_{vert.}$ (W/(m K))	Norm. compr. strength $f_{b,\ min.}$ (MPa)
Homogeneous	Cellular glass	Foamglas (Perinsul)	100–165	0.040–0.050	0.040–0.050	0.4–1.7
	XPS	300–700 L	35–40	0.030–0.040	0.030–0.040	0.3–0.7
	Autoclaved aerated concrete/foam concrete	Ytong	100–650	0.045–0.200	0.045–0.200	0.3–4
Composite	Brick + stone wool	Poroton-MW	550–750	0.070–0.100	/	6–10
	Brick + perlite	Poroton-P	600–900	0.070–0.110	/	4–10
	Lightweight concrete + XPS	Schöck Novomur	700–1000	0.080–0.090	0.190–0.290	6–20
	Nanoconcrete + XPS/PIR	Marmox Thermoblock	300–400	0.044–0.048	0.050–0.075	6.5–10

Note PIR... polyisocyanurate insulation
Source Summarised from catalogues (Marmox 2015; Schöck 2015; Wienerberger 2016; Ytong 2021; Fibran 2020)

3.2.3 Base Insulation Blocks as Parts of Masonry Structures for Bearing Vertical Static Loads

From the aspect of the load-bearing capacity for vertical loads, the only limitation of such base blocks is their compressive strength. The height of a wall and the corresponding vertical loads therefore directly influence the selection of such base blocks. When designing or verifying the load-bearing capacity of walls and masonry structures for vertical loads, we do not take into account the mechanical properties of individual constituent materials (e.g. masonry units or mortar) but of the whole masonry structure, which is considered a homogeneous construction material composed of various materials (Tomaževič 2009). In line with the Eurocode 6 (CEN 2006) rules, the following strength and deformation-related properties are taken into account in the design of masonry structures:

- characteristic compressive strength of masonry (f_k),
- shear strength of masonry (f_v);
- flexural strength of masonry (f_x); and
- relation between stresses and deformations (elastic modulus E and shear modulus G).

All mechanical properties are generally determined through testing according to the standards in the SIST EN 1052 group (Methods of test for masonry). If no data are obtained through research, the characteristic compressive strength of unreinforced masonry, in line with Eurocode 6, can be assessed on the basis of normalised compressive strength of masonry units (f_b) and the compressive strength of mortar (f_m) using the following empirical equation:

$$f_k = K f_b^\alpha f_m^\beta \quad [\text{MPa}], \tag{3.8}$$

whereby K is a constant dependent on the shape and material of the masonry unit and mortar, and α and β are constants dependent on the type of mortar ($\alpha = 0.7$ and $\beta = 0.3$ apply to general purpose mortar and lightweight mortar).

The Eq. (3.8) shows a great influence of the compressive strength of a masonry unit, which is an insulation base block in our case, on the characteristic compressive strength of masonry. Figure 3.14 shows normalised compressive strength of base insulation blocks (f_b) in relation to their density. It may be gathered from Fig. 3.14 that various brick masonry units filled with thermal insulation have the highest strength. Closest to them are various base insulation blocks, which are composites with (nano)concrete, characterised by comparably good load-bearing capacity. When determining loads on masonry, we must bear in mind that base insulation blocks are set at the location where the whole vertical load-bearing structure is fixed on the foundation slab or the basement RC structure. This means that the loads of the whole building are transferred through base insulation blocks to the foundation base. Therefore, certain manufacturers limit the construction of masonry buildings on their base insulation blocks also in non-earthquake-prone areas. Manufacturer

Schöck (2015) permits construction up to four storeys without additional controls of compressive stresses, while the limitations of manufacturer Marmox (2015) are stricter and permit construction up to two storeys with an attic. Additional loads on masonry structures can be expected in the case of seismic action. Therefore, the limitations of base blocks must be explored.

3.3 Details for Preventing Thermal Bridges in Cantilever Structures

At the building connection between the cantilever structure and an outer wall, a significant thermal bridge occurs (due to the penetration of the thermal envelope with the load-bearing structure from reinforced concrete, steel, etc.). In general, such details of balcony structures are included among »+« type connection details and are treated as the most critical thermal bridges in the building envelope. Such discontinuity of the thermal envelope leads to significant heat losses and lower surface temperatures, which may cause much higher heating costs and a high risk of mould at the contact of the cantilever element. To avoid the risk of thermal bridges, special precast load-bearing thermal insulation elements (LBTIEs), which interrupt a thermal bridge at the fixing of the cantilever and can be used for reinforced concrete (RC) slabs and steel and even timber cantilever beams. Although thermal bridges of cantilevers could be limited or eliminated in other ways (e.g. with a separate self-standing balcony structure, a hanging cantilever, etc.), LBTIEs are most frequently used in practice due to their simple use and good energy efficiency.

3.3.1 Precast Element Solutions for Thermal Bridges in Cantilever Structures

Negative effects of thermal bridges between a monolithic RC slab and the balcony were addressed around 1983, when the first solution with precast load-bearing thermal insulation elements (LBTIEs) for RC balcony slabs was designed and is still used in current building practice (Schöck 2020). Today, various solutions (e.g. Halfen (2015), Schöck (2020)) appear on the market for preventing thermal bridges between an RC slab and steel or timber cantilever beams (Fig. 3.15).

Common to all of them is that they enable the continuous thermal envelope, since they are composed of thermal insulation in addition to load-bearing elements (e.g. steel reinforcement). Most versions of LBTIEs are available for the RC structure. They can be used to insulate cantilever balconies, supported balconies, cantilever beams, continuous slabs between a parapet and the ceiling, cantilever balconies with height offsets, etc. The most common elements are used for the insulation of cantilever balconies, which are a composite of thermal insulation made of EPS,

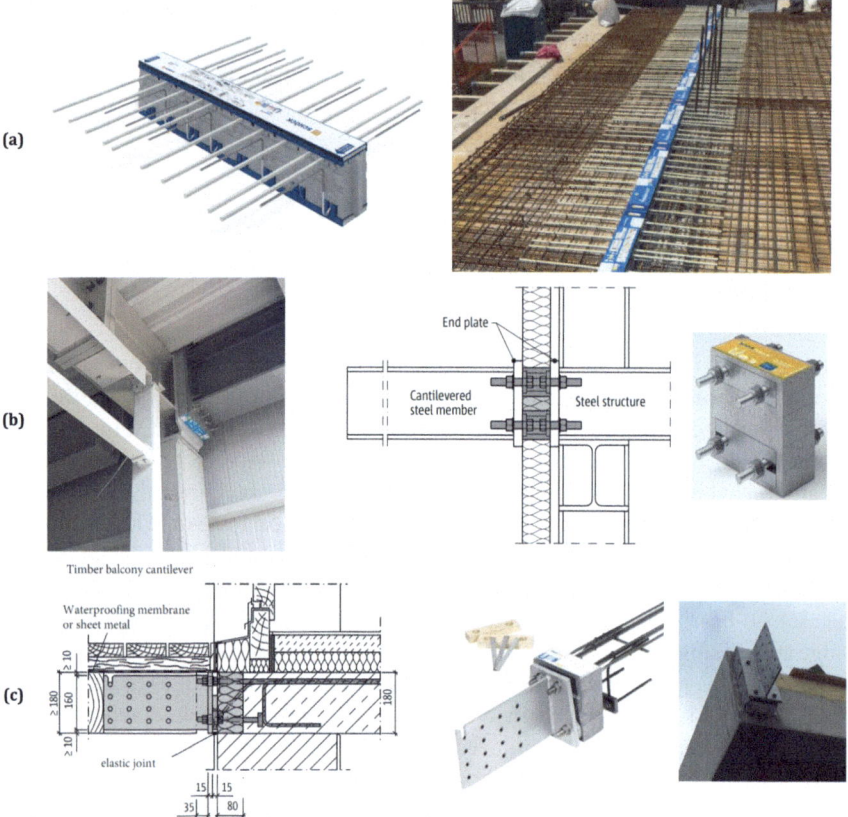

Fig. 3.15 Examples of precast balcony cantilever elements for preventing thermal bridges on the building connection between an outer wall and **a** RC cantilever slab; **b** steel cantilever beams; and **c** timber cantilever beams. *Source* Schöck (2020)

non-corroding steel longitudinal tensile reinforcements, shear reinforcements, and a high-tensile compressive concrete element (Fig. 3.15a). Elements are available for various thicknesses of RC slabs (15–25 cm), various thicknesses of thermal insulation (8–12 cm), and various quantities of longitudinal and shear reinforcements in relation to the required design strength of the cantilever element. LBTIEs are installed and connected with other parts of the reinforcement during the process of reinforcing the slab (Fig. 3.15a).

LBTIEs for steel cantilever beams are a composite of thermal insulation, steel threaded rods, and strong steel end plates (Fig. 3.15b). Solutions for steel cantilever beams are modular and can be applied to various sizes of steel profiles. Attention must be paid to the difference between compressive and tensile elements. Compressive elements primarily transmit pressure but can also transmit some tension and shear forces and prevent the deflection of the rods, while tensile elements are designed exclusively to assume the tensile force. The least common LBTIEs are elements for timber cantilever beams, which are connected with the primary RC load-bearing structure (Fig. 3.15c). The reason lies in the fact that timber elements act as a relatively good thermal insulator (λ_{wood} < 0.35 W/(m K)), making the thermal bridge at the location of the penetration of the thermal envelope smaller than with steel cantilever beams and RC slabs. Nevertheless, most timber cantilever beams in the primary RC structure are fixed with steel anchors, which further increases heat flow, which is why using LBTIEs in certain cases is sensible even with timber cantilever beams.

3.3.2 Comparison of the Environmental and Energy-Efficiency Parameters of an Insulated and Uninsulated Reinforced Concrete Cantilever Slab

From the aspect of environmental and energy-efficiency parameters, the installation of a LBTIE leads to higher surface temperatures and less intensive heat flow (by reducing the thermal bridge) (Goulouti et al. 2014; Ge et al. 2013). Ge and colleagues (2013) analysed the influence of thermal bridges in cantilever structures on energy use in a 26-storey building in boundary climatic conditions in Toronto, Canada. It was established that the surface of these thermal bridges amounts to four percent of the whole surface of the building envelope, leading to an annual increase in energy used for heating of 5–11% in the given building.

This section shows changes in the response of the RC cantilever slab when using a LBTIE in comparison with a thermally uninsulated RC cantilever slab. For this purpose, a study of the temperature range for detail *DB*-01 (uninsulated RC balcony slab) and the cantilever detail was carried out using a LBTIE (Sect. 5.2), whereby the thickness of thermal insulation on the outer wall was changed for each detail. In both details, the selected vertical load-bearing structure was a 25 cm thick brick wall and a 20 cm thick RC interstorey or balcony slab (more details in Sect. 5.2). For all insulated units, the assumed thickness of the LBTIE was 8 cm, and its thermal conductivity was 0.12 W/(m K), which is in line with the current practice. Figure 3.16 shows surface temperatures of the most exposed part of the detail, where heat flow is the highest (on the corner of the outer wall and the ceiling), while Fig. 3.17 further shows the linear thermal transmittance coefficients for the thermal bridge in both models.

The analysis of the temperature range showed that the response of the model using a LBTIE in all structural assemblies had improved, since surface temperatures are higher by up to 5 °C with the same thickness of thermal insulation on the outer wall. Considering the given boundary conditions ($\theta_e = -10$ °C; $\theta_i = 20$ °C and 50-percent relative humidity), Fig. 3.16 reveals that there is a risk of condensation on uninsulated models even with a 10 cm thermal insulation on the outer wall, which still meets the requirements for the thermal transmittance of the outer wall $U_{min.} < 0.28$ W/(m^2 K) in the European regions of continental climate. By installing a LBTIE and with the same thickness of thermal insulation (10 cm), temperatures get close to the desired

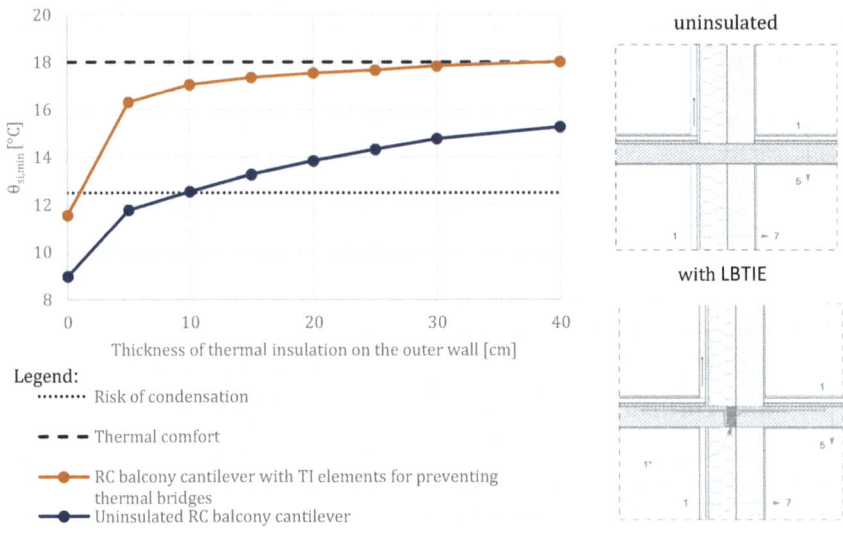

Fig. 3.16 Influence of the LBTIE on the internal surface temperatures of the building connection between the RC cantilever slab, RC interstorey slab, and an outer masonry wall

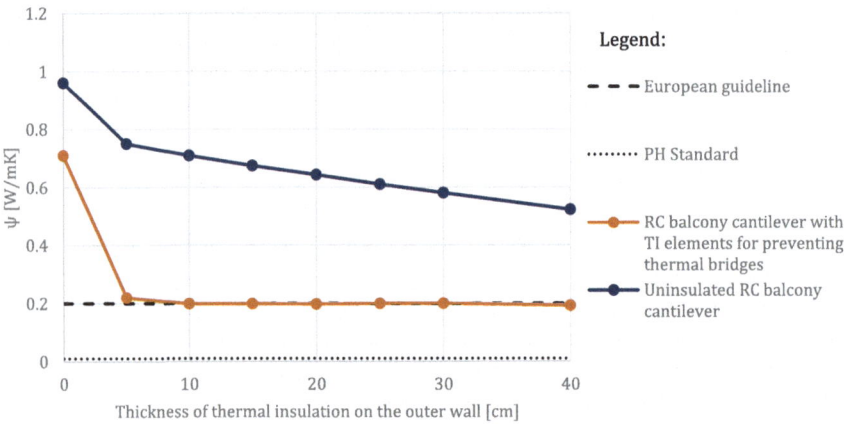

Fig. 3.17 Linear thermal transmittance for the building connection of the thermally uninsulated RC balcony slab compared to the thermally insulated RC balcony slab

thermal comfort (approx. 17 °C), which do not differ significantly from the indoor air temperature ($\theta_i = 20$ °C). It can also be proved that, despite using structural assemblies in accordance with the requirement of the PH standard (thermal insulation on the outer wall is thicker than 20 cm), the uninsulated RC cantilever detail is not appropriate for modern energy-efficient buildings, since surface temperatures are too low to provide thermal comfort. The most considerable influence of a LBTIE is noticeable in structural assemblies with thinner thermal insulation, in which temperature differences between the models are the greatest. On the other hand, the installation of a LBTIE is less effective at a greater thickness of thermal insulation on the outer wall, as higher surface temperatures are not proportional to the uninsulated cantilever model. The results in Fig. 3.17 show that a greater thickness of thermal insulation and the installation of a LBTIE did not completely eliminate the thermal bridge, which must be taken into account in the calculation of heat used for heating in the whole building.

3.3.3 Selection Principles of Precast Load-Bearing Thermal Insulation Elements Subject to Vertical Static Loads

The solution with LBTIEs from the aspect of structural safety is not as good as from the aspect of better energy efficiency, since weakening resulting from the insertion of thermal insulation occurs at the location of the highest loads. Such a weakening may be illustrated through a comparison of the strength of thermal insulation, which is most frequently used for LBTIEs (i.e. EPS and XPS). To make a specific comparison between XPS300 (its nominal compressive strength is 300 kPa) and concrete C30/37 (its characteristic compressive strength is 30 MPa), the strength of the thermal insulation is approximately hundred times lower. For such elements to sustain as high loads as possible, making them comparable with the load-bearing capacity of conventional uninsulated balconies, special load-bearing elements placed in the compression zone of the precast cross-section are added to improve their strength (Keller et al. 2007; Schöck 2020). Compressive load-bearing elements are usually composed of high-performance concretes, e.g. high-quality concretes with micro-steel fibres or synthetic polymers reinforced with glass fibres. In certain precast elements with a high load-bearing capacity, stainless steel studs are also used for compressive elements. The transfer of a force couple due to a negative moment is facilitated through compressive elements on the compressive side, and through tensile reinforcements (RC) or steel rods (in case of steel cantilever beams) on the tensile side. The agreement that a negative moment results in tensile stress on the upper edge of the cross-section is considered.

Manufacturers H-BAU (2021), Halfen (2015), Schöck (2020) provide various precast elements, and their selection depends on the material of the load-bearing structure, the type of the static system, the scale of external actions, and the cantilever length. In addition to thermal insulation, the basic constituents of RC cantilever slabs used for insulating cantilever balconies are tensile and shear reinforcements (rebars at a 45-degree angle) and compressive elements made of high-performance concrete (Fig. 3.18). From the aspect of technical and structural conditions, the design strength of LBTIEs subjected to static vertical loads must be proven sufficient for the combination of actions referred to the ultimate limit state. At the same time,

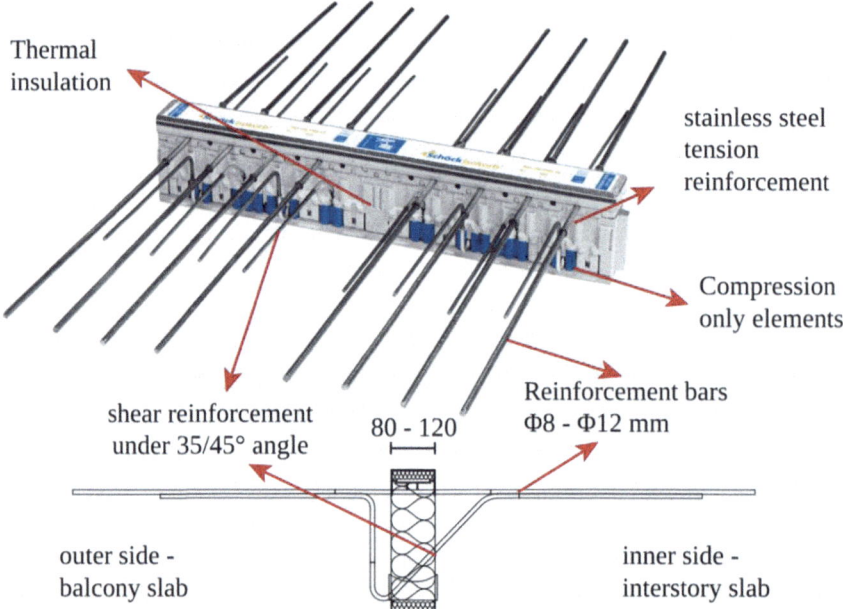

Fig. 3.18 Assembly of the precast LBTIE for RC balcony slabs. *Source* Summarised from catalogue Schöck (2020)

they must show sufficient stiffness not to exceed the maximum cantilever deflection determined for the serviceability limit state.

Ultimate limit state of load-bearing thermal insulation cantilever elements

LBTIEs for cantilever balconies must sustain loads with a negative moment and shear force, which are highest at the fixed end of the cantilever. Since most of the cross-section of these precast elements is composed of thermal insulation, the established equations for designing RC cross-sections cannot be used to determine their resistance to moment and shear loads. Instead, experimental research must be conducted to provide the crucial basis for the acquisition of a certificate of conformity and the sale of such elements on the construction products market. The results of experimental research in relevant scientific literature show that LBTIEs can be used to achieve strength comparable to that of uninsulated cantilevers with a RC cross-section (see, for example, Heidolf and Eligehausen (2013), Keller et al. (2006), Riebel and Keller (2009)).

3.3 Details for Preventing Thermal Bridges in Cantilever Structures

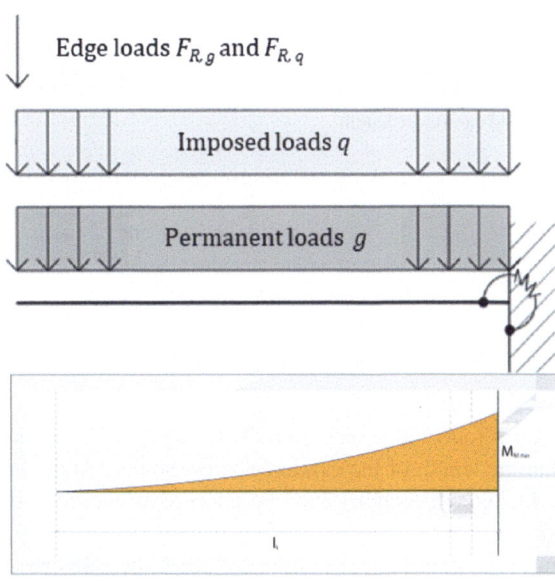

Fig. 3.19 Numerical model of the precast balcony cantilever element with corresponding labels. *Source* Summarised from catalogue Schöck (2020)

Figure 3.19 shows a simple numerical cantilever model for cantilever balconies, which is used to calculate internal forces and design LBTIEs. To prove the load-bearing capacity for the vertical statical load of the model in Fig. 3.19, the Eq. (3.1) can be transformed as follows:

$$m_{Ed} = \left[(\gamma_g g + \gamma_q q)l_k^2/2 + (\gamma_g F_{R,g} + \gamma_q F_{R,q})l_k\right] < m_{Rd}, \qquad (3.9)$$

$$v_{Ed} = \left[(\gamma_g g + \gamma_q q)l_k + \gamma_g F_{R,g} + \gamma_q F_{R,q}\right] < v_{Rd}, \qquad (3.10)$$

According to Eurocode 0 $\gamma_g = 1.35$ and $\gamma_q = 1.50$ are safety factors for permanent and imposed loads in ULS, g is evenly distributed self-weight and dead load, q is evenly distributed imposed loads, $F_{R,g}$ is dead boundary point loads, $F_{R,q}$ is imposed boundary point loads, l_k is systemic cantilever length (Fig. 3.19), m_{Rd} the bending resistance of the LBTIE (experimentally determined value), and v_{Rd} is the shear resistance of the LBTIE (experimentally determined value).

Based on the Eqs. (3.9) and (3.10) LBTIEs are selected according to the data on bending (m_{Rd}) and shear (v_{Rd}) resistance from catalogues based on standardised experimental research.

Cantilever deflection limits

When designing LBTIEs, most attention must be paid to limitations brought on by the maximum cantilever deflection, which increases on account of additional deformation (rotation) of the LBTIE if compared to RC cross-section cantilevers. The total deflection of an insulated cantilever is the sum of the deflection of the concrete cantilever part and deflection due to the deformation of the LBTIE. The

Fig. 3.20 Cantilever end deflection for the serviceability limit state depending on the variable load and cantilever length

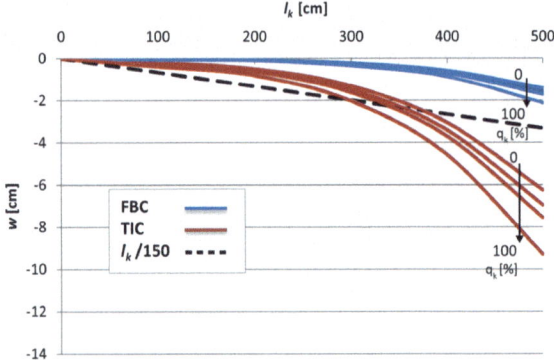

latter is reflected in increased flexibility of the cantilever, which in most cases means that the length of the cantilever must be limited or certain structural measures (e.g. reducing self-weight) must be adopted to facilitate the use of such elements. The comparison of deflections of insulated (TIC) and uninsulated (i.e. fixed base, FBC) models (comparable by static height of the slab cross section) in the serviceability limit state due to vertical static loads is shown in Fig. 3.20 where LBTIEs from practice were analysed (Azinović et al. 2015a). The weight of the concrete slab, which differs according to the thickness of the slab ($25\frac{kN}{m^3} H_k$ (m)), and the remaining dead load ($1.5\,kN/m^2$) were taken into account in the calculation of the self-weight and dead load of the models (g_k). Imposed load was added as a point force at the end of the cantilever ($Q_k = 1.0\,kN/m$) and as a linear continuously distributed load (q_k), which is changed from 0 to 5 kN/m².

If the results are compared to the maximum permissible deflection of the cantilever (SLS: $w_{\max} = l_k/150$) determined in Eurocode 0, it can be noticed that in the analysed example, the FBC model met the conditions regardless of the length of the model. If the limit deflection of the cantilever is exceeded, the appearance and general usefulness of the structure could deteriorate. Nevertheless, in most cases, cantilever is still not damaged at such a displacement (remains in elastic state). On the other hand, the analysis showed that the TIC models display the highest deflection values, and the difference with the comparable FBC models can be five times and more. Taking into account the limit of critical deflection in the serviceability limit state ($l_k/150$), the approximate maximum length to which LBTIEs can still be used can be determined. As evident from the graph in Fig. 3.20, the critical deflection of the analysed TIC models with the highest imposed load (q) occurs at the length of 300 cm. At the lowest imposed load, the analysed cantilevers can be slightly longer, as the limit stands at approx. 350 cm (for selected case study examples). The main limitation of LBTIEs is not their insufficient strength, but their excessive deflection, which may be exceeded already for vertical static loads. Our analyses showed that conventional LBTIEs available on the market can be used for RC cantilevers shorter than 300 cm, whereas the use of such elements for longer cantilevers is not recommended.

3.3 Details for Preventing Thermal Bridges in Cantilever Structures

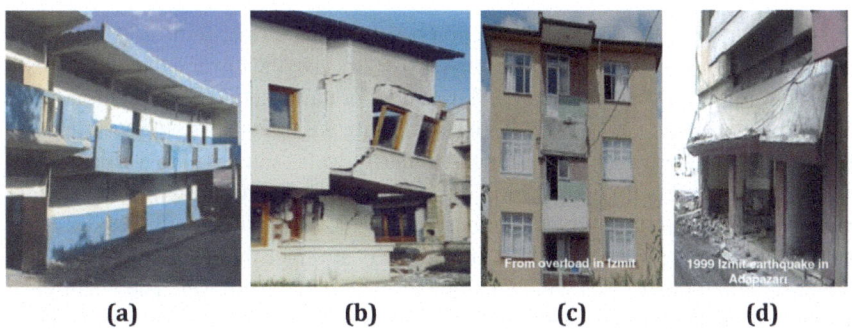

(a) (b) (c) (d)

Fig. 3.21 Structural damage to cantilevers caused by the Izmit earthquake (Turkey, 1999). *Source* Dogan et al. (2007)

The influence of LBTIEs on the seismic response of cantilever structures are shown in more detail in Azinović et al. (2014a, 2015a). In addition to the technical and structural conditions which must be considered by designers for vertical static loads, certain specifics must also be taken into account for cantilever structures in earthquake-prone areas. These specifics include a change in vibration (extended fundamental period) in the vertical direction for insulated cantilevers (the TIC model), which is very important when addressing their dynamic response. Extended fundamental period is a result of greater flexibility of insulated models (TIC), which also increases cantilever deflection. The length of the fundamental period affects the determination of seismic forces and provides information of the model stiffness. Analyses have shown that the fundamental period can be extended by as much as 2.5 times.

In addition to the change in the fundamental period of insulated models of cantilever structures, critical response mechanisms under seismic action must also be known (more in Fig. 3.21). Similar to vertical static loads, exceeding limit deflection (Fig. 3.21a) is pointed out as the first limit state. Since LBTIEs are very flexible, great deflections of a cantilever can occur before serious damage of the LBTIEs. The scale of the critical deflection of a cantilever under seismic action is not prescribed in regulations, and may considerably differ, depending on whether also other (non-)structural elements (e.g. conservatory, glass railing, etc.) must be protected. Designers and investors must prevent critical deflections if the risk that an earthquake could cause too much damage to a cantilever or other secondary elements (see, for example, Fig. 3.21b) is too high. It must be pointed out that an earthquake is a momentary and short-term load, which means that, in most cases, deflection is exceeded momentarily, which, however, can still damage secondary (fragile) elements.

Another limit state could be, in addition to excessive deflection, the exceeded bending resistance of the precast cross-section (Fig. 3.21c, d). In this case, the upper side of the concrete slab or the insulation elements are visibly damaged. Such a critical state can develop particularly in the case of an earthquake on a fully loaded cantilever (loaded with the whole assumed imposed load). That is when the ultimate tensile stress in the upper part of the cross-section (longitudinal tensile reinforcement in the case of a RC cross-section) and the compressive stress in the lower edge of the cross-section (compressive elements in the TIC model or the concrete cross-section in the FBC model) are exceeded. Such a limit state is only possible in extreme earthquakes, since cantilever elements are designed primarily to be safe at high vertical static loads, which are usually a governing design case and therefore high safety factors are applied. The cantilever uplift is pointed out as the last critical state. It occurs if an earthquake in the opposite direction of gravity is stronger than vertical static loads (self-weight and dead load) which act downward on the cantilever. The occurrence of such a borderline case is more probable in the event of seismic action on a less loaded cantilever (without imposed load). The cantilever uplift can lead to exceeded tensile stress in the lower part of the cross-section and exceeded compressive stress in the upper part. A momentary cantilever uplift during an earthquake is not critical if the lower part of its cross-section is strong enough to withstand tensile stress, which is in practice not common due to lacking steel reinforcement at the lower end of LBTIEs. Otherwise, the lower part of the cantilever will be more seriously damaged due to exceeded tensile strength.

The cantilever detail can change the seismic response of a building only locally and bring about only a local collapse of the cantilever, while its impact on the total safety of the primary load-bearing structure is low. Certain deterioration in the load-bearing capacity of the »+« type connection between the RC interstorey and balcony slabs and an outer wall is possible but highly unlikely. In addition, such local damage to cantilevers is admissible in view of the established principle of the capacity design method, since vertical load-bearing elements (e.g. outer walls, columns) are primarily protected when designing earthquake-resistant structures. Nevertheless, if a cantilever collapsed onto the street, it could injure building users during evacuation in the event of a strong earthquake (Costa et al. 2020; Koren and Rus 2021; Santarelli et al. 2018). The latter turned out to be very problematic in older masonry buildings in the recent earthquake in Zagreb (Stepinac et al. 2021).

3.4 Building Connection Detail Between the Roof and Outer Wall

Outer walls and roofs bordering conditioned spaces are locations in the building envelope with high transmission losses. Therefore, requirements for the thermal transmittance of such structural assemblies in modern energy-efficient buildings are very strict, contributing to a better thermal envelope and a lower use of energy for heating or cooling. The elimination of thermal bridges on the building connection between the roof and an outer wall also fosters better use of energy (see Fig. 3.22). The requirement for continuous thermal insulation on this connection could in certain cases lead to a poorer structural contact, poorer fixing of the roof structure to vertical load-bearing elements, more difficult roof anchorage, etc. Most roof connection

3.4 Building Connection Detail Between the Roof and Outer Wall

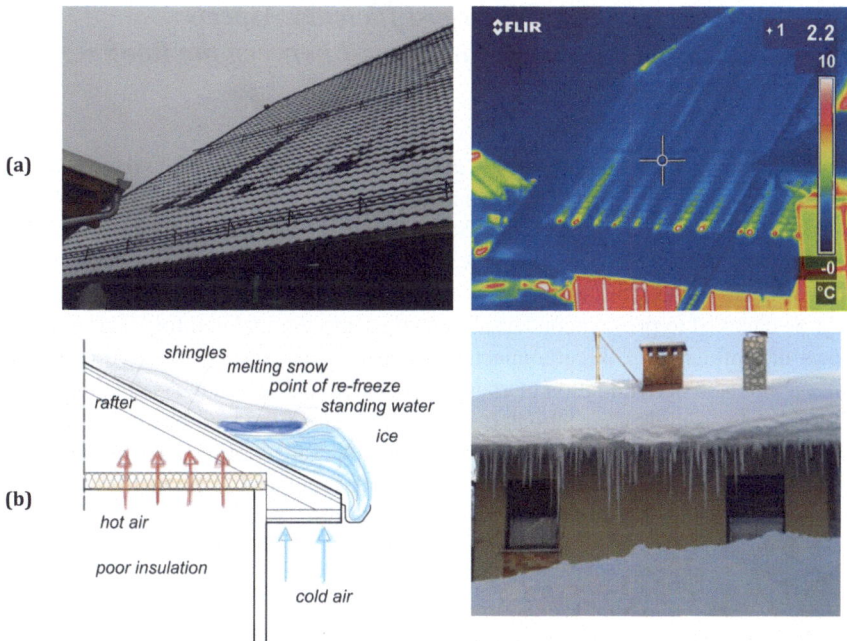

Fig. 3.22 Examples of thermal bridges on roof structures and connections between the roof and an outer/inner wall. *Source* Building and Civil Engineering Institute ZRMK (author: M. Tomšič) and B. Azinović

details are »L« type (connections of a flat roof without parapets or a pitched roof without an overhang) and »T« type (flat roofs with parapets or overhangs, pitched roofs with overhanging eaves) details. In the case of »L« type details, thermal bridges occur particularly in structural assemblies with thermal insulation on the internal side or in the core of the load-bearing structure (Fig. 3.5). In the case of »T« type details, most thermal bridges stem from connections of various overhangs, parapets or overhanging eaves, which are not thermally insulated.

Certain examples of poor practice when thermal bridges occur in roof structures or at the connection between the roof and an outer or inner wall are shown in Fig. 3.22. It is clearly evident from Fig. 3.22 how the pattern of melting snow reveals thermal bridges. Indicators of unsuitable structural assemblies and a poor connection between the roof and an outer wall can also be icicles (Fig. 3.22c). The effects of poor connections are presented in more detail in Sect. 3.4.1, in which a specific case of a reinforced concrete (RC) flat roof with a parapet is discussed from the environmental and energy-efficiency aspect. Finally, technical and structural requirements are provided, which must be met by the roof structure and the connection between the roof and an outer wall for vertical static loads as well as for seismic actions.

3.4.1 Environmental and Energy-Efficiency Aspects of the Building Connection Detail Between the Roof and an Outer Wall

Taking into account the environmental and energy-efficiency aspect in the design of the connection detail between the roof and an outer wall is crucial, which is supported by examples of poor practice in Fig. 3.22. To design the detail, thermal transmittance of both structural assemblies and a low coefficient of linear transmission losses for thermal bridges must be determined, and surface temperatures around the detail must be checked to prevent condensation and ensure thermal comfort. This section shows the influence of three connection solutions between the roof and an outer wall on the local determination of energy parameters of the detail. As combinations of structural assembly connections (various materials of the load-bearing structure, thermal insulation position, flat or pitched roofs, etc.) are endless, the study focuses on the specific example of the detail of the RC flat roof with a parapet (Fig. 3.23). More solutions of different connection details between the roof and an outer wall are given in the Appendix.

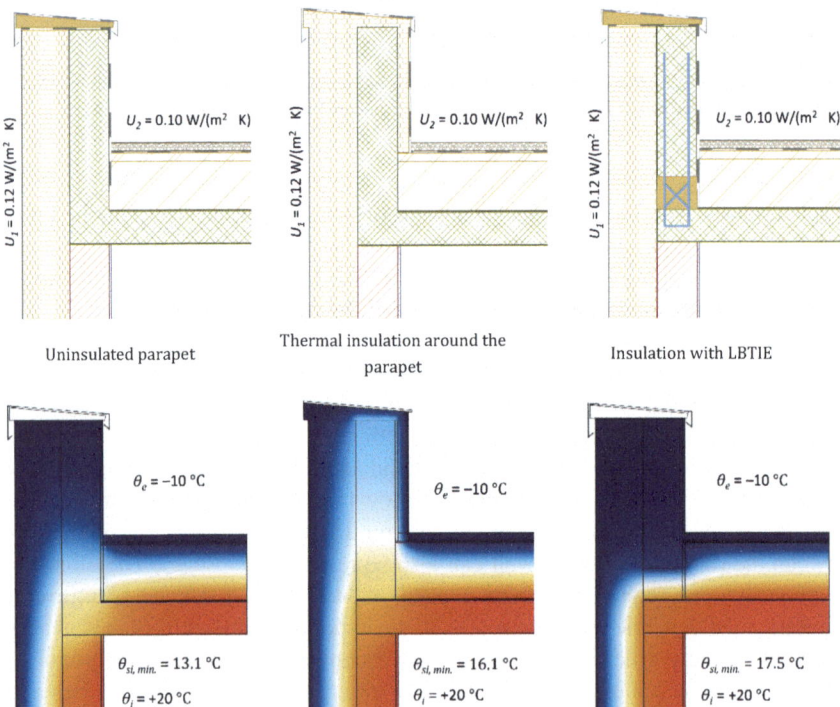

Fig. 3.23 Internal surface temperatures of the building connection between the RC flat roof, parapet, and an outer wall

3.4 Building Connection Detail Between the Roof and Outer Wall

Figure 3.23 reveals three connection solutions between the flat roof with a parapet with the same composition of the two basic structural assemblies: an outer masonry wall (thermal transmittance of the assembly—$U_1 = 0.12\,\text{W}/(\text{m}^2\,\text{K})$) and the RC flat roof ($U_2 = 0.10\,\text{W}/(\text{m}^2\,\text{K})$). In the first case, the load-bearing structure is not interrupted, which leads to a thermal bridge in such a »T« type detail. In the second case, such a thermal bridge is reduced with continuous thermal insulation around the parapet. In practice, the thickness of additional thermal insulation on the internal side of a parapet does not exceed five centimetres, which was taken into account in the analysis. In the third case, the load-bearing thermal insulation element (LBTIE) that interrupts a thermal bridge at the fixing point of the parapet (similar to LBTIEs for RC balcony slabs).

The temperature range analysis results have shown that the detail with an uninsulated parapet is not to be used in energy-efficient buildings, as surface temperatures come close to the condensation point despite the use of structural assemblies with a very low thermal transmittance ($U < 0.12\,\text{W}/(\text{m}\,\text{K})$). The same conclusion can be made on the basis of Fig. 3.24, which shows the linear thermal transmittance coefficient for the thermal bridge at the analysed building connection. Among all the analysed details, heat losses of the uninsulated parapet detail are the highest and stand at over 0.40 W/(m K), regardless of the thermal transmittance of both connected structural assemblies. As expected, details with continuous thermal insulation around the parapet exhibit a better response. Such details could be used in energy-efficient buildings if certain types of weakening were taken into account. In the analysed case, surface temperatures are higher than 16 °C, which provides significant protection against condensation, approaching the desired temperatures for thermal comfort. However, heat losses, which can amount to over 0.20 W/(m K), must be taken into account for the insulated detail around the parapet. The reason for relatively high heat losses despite continuous thermal insulation around the parapet can most simply

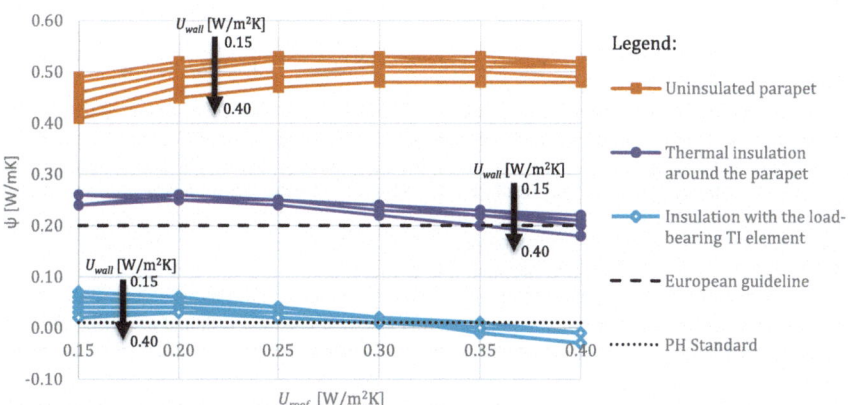

Fig. 3.24 Linear thermal transmittance for different solutions of the building connection between the RC flat roof, a parapet and an outer masonry wall. *Source* Summarised from EnergieSchweiz (2002), pp. 35–37

be illustrated with the detail's temperature range (Fig. 3.24). The latter shows that thermal insulation stops the cold outdoor air, but heat losses still occur due to the heating of the whole RC parapet with a high thermal conductivity. The best response can be expected from RC parapets with LBTIEs, where temperatures on the inner surface are higher than 17.5 °C and losses resulting from a thermal bridge are limited ($\psi < 0.10$ W/(m K)).

In the analysed three cases of the parapet and flat roof detail, the structural contact between the load-bearing outer masonry wall and the RC flat roof does not change, meaning that there are few or no negative impacts on the structural safety of the building. Nevertheless, the load-bearing capacity of the parapet could be locally reduced in certain cases due to requirements for continuous thermal insulation. Certain details of parapets with LBTIEs used in non-earthquake-prone areas are anchored on the primary load-bearing structure with asymmetrical reinforcement or asymmetrical fasteners. In the case of parapets with a great mass (e.g. concrete parapets) and a large surface, the latter could mean increased risk of the parapet collapse, putting the life of people near the building at risk and causing additional material damage.

References

Allotey N, El Naggar MH (2008) An investigation into the Winkler modeling of the cyclic response of rigid footings. Soil Dyn Earthq Eng 28:44–57

Anand V, Kumar SS (2018) Seismic soil-structure interaction: a state-of-the-art review. In: Structures. Elsevier, pp 317–326

Anastasopoulos I, Kontoroupi T (2014) Simplified approximate method for analysis of rocking systems accounting for soil inelasticity and foundation uplifting. Soil Dyn Earthq Eng 56:28–43

Arioz O, Kilinc K, Karasu B, Kaya G, Arslan G, Tuncan M, Tuncan A, Korkut M, Kivrak S (2008) A preliminary research on the properties of lightweight expanded clay aggregate. J Aust Ceram Soc 44:23–30

Athanasopoulos GA, Pelekis PC, Xenaki VC (1999) Dynamic properties of EPS geofoam: an experimental investigation. Geosynth Int 6:171–194

Auad G, Almazán JL (2021) Lateral impact resilient double concave friction pendulum (LIR-DCFP) bearing: formulation, parametric study of the slider and three-dimensional numerical example. Eng Struct 233:111892

AURE (2003) Materials for thermal insulation (In Slovene) [Online]. http://www.aure.gov.si/eknjiznica/IL_2-03.PDF. Accessed 15 March 2013

Azinović B, Koren D, Kilar V (2014a) Seismic safety of the precast balcony cantilever elements for prevention of thermal bridges. Architec Res 2014:25–33

Azinović B, Koren D, Kilar V (2014b) Seismic safety of low-energy buildings founded on a thermal insulation layer. In: A joint event of the 15th European conference on earthquake engineering and 34 general assembly of the European seismological commission, 24–29 August 2014, Istanbul, Turkey

Azinović B, Koren D, Kilar V (2014c) The seismic response of low-energy buildings founded on a thermal insulation layer—a parametric study. Eng Struct 81:398–411

Azinović B, Kilar V, Koren D (2015a) Erdbebensicherheit vorgefertigter wärmegedämmter Stahlbeton-Konsolenelemente. Bauingenieur 90:489–499

References

Azinović B, Koren D, Kilar V (2015b) Sliding isolation system for controlled response of passive houses founded on thermal insulation. In: 14th World conference on seismic isolation, energy dissipation and active vibration control of structures. Anti-Seismic Systems International Society (ASSISi), San Diego, USA

Azinović B, Kilar V, Koren D (2016) Energy-efficient solution for the foundation of passive houses in earthquake-prone regions. Eng Struct 112:133–145

Banović I, Radnič J, Grgić N (2020) Foundation size effect on the efficiency of seismic base isolation using a layer of stone pebbles. Earthq Struct 19:103–117

Becker TC, Mahin SA (2012) Experimental and analytical study of the bi-directional behavior of the triple friction pendulum isolator. Earthquake Eng Struct Dynam 41:355–373

Brandis A, Kraus I, Petrovčič S (2021) Simplified numerical analysis of soil-structure systems subjected to monotonically increasing lateral load. Appl Sci 11:4219

Bunge F, Merkel H (2011) Development, testing and application of extruded polystyrene foam (XPS) insulation with improved thermal properties. Bauphysik 33:67–72

Cavalieri F, Correia AA, Crowley H, Pinho R (2020) Dynamic soil-structure interaction models for fragility characterisation of buildings with shallow foundations. Soil Dyn Earthquake Eng 132:106004

CEN (2004a) Eurocode 0: basis of structural design. EN 1990:2004

CEN (2004b) Eurocode 1: actions on structures. EN 1991-1:2004

CEN (2005a) Eurocode 2: design of concrete structures—Part 1-1: general rules and rules for buildings. EN 1992:2005

CEN (2005b) Eurocode 7: geotechnical design—Part 1: general rules. EN 1997-1:2005

CEN (2006) Eurocode 6—design of masonry structures—Part 1-1: general rules for reinforced and unreinforced masonry structures. EN 1996-1:2006

CEN (2013a) Thermal insulating products for building applications—determination of compression behaviour. EN 826:2013

CEN (2013b) Thermal insulating products for building applications—determination of compressive creep. EN 1606:2013

CEN (2013c) Thermal insulation products for buildings—factory made cellular glass (CG) products—specification. EN 13167:2013

CEN (2013d) Thermal insulation products for buildings—factory made mineral wool (MW) products—specification. EN 13162:2013

CEN (2013e) Thermal insulation products for buildings—factory made products of expanded polystyrene (EPS)—specification. EN 13163:2013

CEN (2013f) Thermal insulation products for buildings—factory made products of extruded polystyrene foam (XPS)—specification. EN 13164:2013

CEN (2013g) Thermal insulation products for buildings—factory made rigid polyurethane foam (PUR) products—specification. EN 13165:2013

Chopra AK, Yim SC-S (1985) Simplified earthquake analysis of structures with foundation uplift. J Struct Eng 111:906–930

Christopoulos C, Filiatrault A, Bertero VV (2006) Principles of passive supplemental damping and seismic isolation. IUSS Press

Chung LL, Kao PS, Yang CY, Wu LY, Chen HM (2015) Optimal frictional coefficient of structural isolation system. J Vib Control 21:525–538

Costa C, Figueiredo R, Silva V, Bazzurro P (2020) Application of open tools and datasets to probabilistic modeling of road traffic disruptions due to earthquake damage. Earthquake Eng Struct Dynam 49:1236–1255

Demirboğa R, Gül R (2003) Thermal conductivity and compressive strength of expanded perlite aggregate concrete with mineral admixtures. Energy Build 35:1155–1159

Dhadse GD, Ramtekkar G, Bhatt G (2020) Finite element modeling of soil structure interaction system with interface: a review. Arch Comput Meth Eng:1–18

Diascorn N, Calas S, Sallée H, Achard P, Rigacci A (2015) Polyurethane aerogels synthesis for thermal insulation—textural, thermal and mechanical properties. J Supercrit Fluids 106:76–84

DIBT (2013) Extrudierte Polystyrol-Hartschaumplatten 'FIBRANxps 300-L' und 'FIBRANxps 500-L' für die Anwendung als lastabtragende Wärmedämmung unter Gründungsplatten. Zulassung Z-23.34-1807

Dogan M, Unluoglu E, Ozbasaran H (2007) Earthquake failures of cantilever projections buildings. Eng Fail Anal 14:1458–1465

Dupre (2015) Vermiculite loose fill insulation [Online]. Available: http://www.dupreminerals.com/en/vermiculite/applications/insulation/cavity. Accessed 14 Dec 2015

Dutta SC, Bhattacharya K, Roy R (2004) Response of low-rise buildings under seismic ground excitation incorporating soil–structure interaction. Soil Dyn Earthq Eng 24:893–914

Elfoam (2020) Technical data Elfoam boards [Online]. Available: http://www.elliottfoam.com/tech.html. Accessed 5 Dec 2020

EnergieSchweiz (2002) Wärmebrückenkatalog. BBL, Bern, Switzerland

EUMEPES (2014) EPS White Book, EUMEPS background information on standardisation of EPS [Online]. Available: http://eumeps.org/show.php?ID=4522&psid=xwctaave. Accessed 17 Dec 2015

Fadi F, Constantinou MC (2010) Evaluation of simplified methods of analysis for structures with triple friction pendulum isolators. Earthquake Eng Struct Dynam 39:5–22

Fibran (2020) Product catalogues: thermal insulation from extruded polystyrene FIBRANxps [Online]. Available: https://fibran.de/wp-content/uploads/sites/12/2020/01/0100_FIBRAN_Product_Catalogue.pdf. Accessed 5 Jan 2020

Foamglas (2015) Erdberührte Dämmsysteme (Perimeterdämmung) Referenzen [Online]. Available: http://de.foamglas.com/de/waermedaemmung/referenzen/#1-1. Accessed 15 Dec 2015

Foamglas (2020) FOAMGLAS insulation for the building envelope [Online]. Available: https://www.foamglas.com/-/media/project/foamglas/public/corporate/foamglascom/files/brochures/building-envelope/foamglas-insulation-for-the-building-envelope-uk.pdf?la=en-gb. Accessed 14 March 2021

Forcellini D (2020) Assessment of geotechnical seismic isolation (GSI) as a mitigation technique for seismic hazard events. Geosciences 10:222

Gazetas G, Anastasopoulos I, Apostolou M (2007) Shallow and deep foundations under fault rupture or strong seismic shaking. In: Pitilakis KD (ed) Earthquake geotechnical engineering. Springer

Gazetas G, Anastasopoulos I, Adamidis O, Kontoroupi T (2013) Nonlinear rocking stiffness of foundations. Soil Dyn Earthq Eng 47:83–91

Ge H, McClung VR, Zhang SS (2013) Impact of balcony thermal bridges on the overall thermal performance of multi-unit residential buildings: a case study. Energy Build 60:163–173

Gelagoti F, Kourkoulis R, Anastasopoulos I, Gazetas G (2012a) Rocking isolation of low-rise frame structures founded on isolated footings. Earthquake Eng Struct Dynam 41:1177–1197

Gelagoti F, Kourkoulis R, Anastasopoulos I, Gazetas G (2012b) Rocking-isolated frame structures: margins of safety against toppling collapse and simplified design approach. Soil Dyn Earthq Eng 32:87–102

GLAPOR (2020) GLAPOR celullar glass gravel [Online]. Available: https://www.glapor.de/en/produkte/cellular-glass-gravel/. Accessed 14 Dec 2020

Gnip IJ, Vaitkus SI, Kersulis VI, Veyelis SA (2007) Deformability of expanded polystyrene under short-term compression. Mech Compos Mater 43:433–444

Gnip IY, Vaitkus S, Keršulis V, Vėjelis S (2010) Experiments for the long-term prediction of creep strain of expanded polystyrene under compressive stress. Polym Testing 29:693–700

Gnip IY, Vaitkus S, Keršulis V, Vėjelis S (2011) Analytical description of the creep of expanded polystyrene (EPS) under long-term compressive loading. Polym Testing 30:493–500

Goulouti K, de Castro J, Vassilopoulos AP, Keller T (2014) Thermal performance evaluation of fiber-reinforced polymer thermal breaks for balcony connections. Energy Build 70:365–371

Halfen (2015) Halfen HIT insulated connection—technical product information [Online]. Germany. Available: http://www.halfen.com/en/769/products/reinforcement-systems/halfen-hit-insulated-connection/introduction/. Accessed 8 Jan 2021

References

H-BAU (2021) Thermal insulation elements for balconies and thermally isolated external components. Germany

Heidolf T, Eligehausen R (2013) Design concept for load bearing thermal insulation elements with compression shear bearings. Beton-Und Stahlbetonbau 108:179–187

Hess Perlite (2015) Underslab insulation using perlite in bags [Online]. Available: http://www.hessperlite.com/perlite-applications.html. Accessed 14 Dec 2015

IBO (2008) Details for passive houses: a catalogue of ecologically rated constructions, 3rd edn. Springer-Verlag, Vienna

Janetti MB, Plaz T, Ochs F, Klesnil O, Feist W (2015) Thermal conductivity of foam glass gravels: a comparison between experimental data and numerical results. Energy Procedia 78:3258–3263

Jarernprasert S, Bazan-Zurita E, Bielak J (2013) Seismic soil-structure interaction response of inelastic structures. Soil Dyn Earthq Eng 47:132–143

JUBHome (2016) Instructions for design of JUBHome BASE [in Slovene] [Online]. Available: http://www.jub.si/jubhome-hise/nasveti/navodila-za-projektiranje-jubhome-base. Accessed 5 Dec 2020

Kan A, Demirboğa R (2009) A novel material for lightweight concrete production. Cement Concr Compos 31:489–495

Keller T, Riebel F, Zhou AX (2006) Multifunctional hybrid GFRP/steel joint for concrete slab structures. J Compos Constr 10:550–560

Keller T, Riebel F, Zhou AX (2007) Multifunctional all-GFRP joint for concrete slab structures. Constr Build Mater 21:1206–1217

Kilar V, Koren D (2009) Seismic behaviour of asymmetric base isolated structures with various distributions of isolators. Eng Struct 31:910–921

Kilar V, Koren D, Zbašnik-Senegačnik M (2013) Seismic behaviour of buildings founded on thermal insulation layer. Građevinar 65:423–433

Kilar V, Azinović B, Koren D (2014a) Energy efficient construction and the seismic resistance of passive houses. In: WASET international conference on sustainable architecture and urban design engineering, Dubai, UAE

Kilar V, Koren D, Bokan-Bosiljkov V (2014b) Evaluation of the performance of extruded polystyrene boards—implications for their application in earthquake engineering. Polym Testing 40:234–244

Kilar V, Azinović B, Koren D (2016) Foundation concept for passive houses in seismic areas (In Slovene). Gradbeni Vestnik 65:59–70

Koren D, Rus K (2021) Assessment of a city's performance under different earthquake scenarios. In: Lakušić S (ed) 1CroCEE. Zagreb, Croatia

Kourkoulis R, Anastasopoulos I, Gelagoti F, Kokkali P (2012a) Dimensional analysis of SDOF systems rocking on inelastic soil. J Earthquake Eng 16:995–1022

Kourkoulis R, Gelagoti F, Anastasopoulos I (2012b) Rocking isolation of frames on isolated footings: design insights and limitations. J Earthquake Eng 16:374–400

Krause P, Nowoświat A, Pawłowski K (2020) The impact of internal insulation on heat transport through the wall: case study. Appl Sci 10:7484

Laterlite (2007) Geotechnical engineering—lightweight solutions with expanded clay Laterlite [Online]. Available: http://www.laterlite.com/literature/. Accessed 14 June 2015

Lukić I, Premrov M, Passer A, Leskovar VŽ (2021) Embodied energy and GHG emissions of residential multi-storey timber buildings by height - A case with structural connectors and mechanical fasteners. Energy Build 111387

Lou M, Wang H, Chen X, Zhai Y (2011) Structure–soil–structure interaction: literature review. Soil Dyn Earthq Eng 31:1724–1731

Lu LY, Lee TY, Juang SY, Yeh SW (2013) Polynomial friction pendulum isolators (PFPIs) for building floor isolation: an experimental and theoretical study. Eng Struct 56:970–982

Lyons A (2014) Materials for architects and builders. Routledge

Mahmoud S, Austrell P-E, Jankowski R (2012) Simulation of the response of base-isolated buildings under earthquake excitations considering soil flexibility. Earthq Eng Eng Vib 11:359–374

Maleki S, Ahmadi F (2011) Using expanded polystyrene as a seismic energy dissipation device. J Vib Control 17:1481–1497

Manzano-Agugliaro F, Montoya FG, Sabio-Ortega A, García-Cruz A (2015) Review of bioclimatic architecture strategies for achieving thermal comfort. Renew Sustain Energy Rev 49:736–755

Marincioni V, May N, Altamirano-Medina H (2015) Parametric study on the impact of thermal bridges on the heat loss of internally insulated buildings. Energy Procedia 78:889–894

Marmox (2015) The ideal solution for thermal bridges—Marmox thermoblock [Online]. Available: http://www.marmox.co.uk/products/thermoblock. Accessed 27 Dec 2015

Méar F, Yot P, Viennois R, Ribes M (2007) Mechanical behaviour and thermal and electrical properties of foam glass. Ceram Int 33:543–550

Merkel H (2004) Determination of long-term mechanical properties for thermal insulation under foundations. In: Buildings conference, pp 1–7

Modic M (2009) Technological process of making EPS insulation boards (In Slovene). Master Thesis, University of Maribor

Mojsilović N, Simundic G, Page A (2010) Masonry wallettes with damp-proof course membrane subjected to cyclic shear: an experimental study. Constr Build Mater 24:2135–2144

MOP (2010) Technical guideline for construction TSG-1-004: 2010, Energy efficiency (In Slovene) [Online]. Slovenian Ministry of the Environment and Spatial Planning. Available: http://www.arhiv.mop.gov.si/fileadmin/mop.gov.si/pageuploads/zakonodaja/prostor/graditev/TSG-01-004_2010.pdf. Accessed 22 July 2018

Mylonakis G, Gazetas G (2000) Seismic soil-structure interaction: beneficial or detrimental? J Earthquake Eng 04:277–301

Naeim F, Kelly JM (1999) Design of seismic isolated structures: from theory to practice. John Wiley & Sons

Nakhaei M, Ali Ghannad M (2008) The effect of soil–structure interaction on damage index of buildings. Eng Struct 30:1491–1499

Nanda RP, Agarwal P, Shrikhande M (2012) Base isolation by geosynthetic for brick masonry buildings. J Vib Control 18:903–910

Ozguven A, Gunduz L (2012) Examination of effective parameters for the production of expanded clay aggregate. Cement Concr Compos 34:781–787

Panchal VR, Jangid RS (2009) Seismic response of structures with variable friction pendulum system. J Earthquake Eng 13:193–216

Papadopoulos AM (2005) State of the art in thermal insulation materials and aims for future developments. Energy Build 37:77–86

Passivhaus Institut (2012) Certification criteria for residential passive house buildings [Online]. Darmstadt, Germany. Available: http://www.passiv.de/downloads/03_certfication_criteria_residential_en.pdf. Accessed 18 August 2013

Premrov M, Leskovar VŽ, Mihalič K (2015) Influence of the building shape on the energy performance of timber-glass buildings in different climatic conditions. Energy:1–11

Ramsteiner F, Fell N, Forster S (2001) Testing the deformation behaviour of polymer foams. Polym Testing 20:661–670

Raychowdhury P (2011) Seismic response of low-rise steel moment-resisting frame (SMRF) buildings incorporating nonlinear soil–structure interaction (SSI). Eng Struct 33:958–967

Raychowdhury P, Hutchinson TC (2009) Performance evaluation of a nonlinear Winkler-based shallow foundation model using centrifuge test results. Earthquake Eng Struct Dynam 38:679–698

Riebel F, Keller T (2009) Structural behavior of multifunctional GFRP joints for concrete structures. Constr Build Mater 23:1620–1627

Rockwool (2020) Thermal insulation slab Rockwool [Online]. Available: https://www.rockwool.com/north-america/products-and-applications/floor-insulation/under-slab-insulation/. Accessed 14 March 2021

Roy R, Chandra Dutta S (2010) Inelastic seismic demand of low-rise buildings with soil-flexibility. Int J Non-linear Mech 45:419–432

References

Santarelli S, Bernardini G, Quagliarini E (2018) Earthquake building debris estimation in historic city centres: From real world data to experimental-based criteria. Int J Disaster Risk Reduc 31:281–291

Schöck (2015) Technical Information Novomur®/Novomur® light [Online]. Available: https://www.schoeck.de/de/produktloesungen/novomur--38. Accessed 4 Dec 2015

Schöck (2020) Technical information Schöck Isokorb® [Online]. Available: https://www.schoeck.com/en-gb/isokorb. Accessed 4 Dec 2020

Sengul O, Azizi S, Karaosmanoglu F, Tasdemir MA (2011) Effect of expanded perlite on the mechanical properties and thermal conductivity of lightweight concrete. Energy Build 43:671–676

Skinner RI, Robinson WH, Mcverry GH (1993) An introduction to seismic isolation

Stepinac M, Lourenço PB, Atalić J, Kišiček T, Uroš M, Baniček M, Šavor Novak M (2021) Damage classification of residential buildings in historical downtown after the ML5.5 earthquake in Zagreb, Croatia in 2020. Int J Disaster Risk Reduc 56:102140

Timsina S, Calvi PM (2021) Variable friction base isolation systems: Seismic performance and preliminary design. J Earthquake Eng 25:93–116

Tomaževič M (2009) Earthquake resistant masonry buildings (In Slovene), Ljubljana, Tehnis

Tsai CS, Su HC, Chiang TC (2014) Equivalent series system to model a multiple friction pendulum system with numerous sliding interfaces for seismic analyses. Earthq Eng Eng Vib 13:85–99

Vejelis S, Gnip I, Vaitkus S, Kersulis V (2008) Shear strength and modulus of elasticity of expanded polystyrene (EPS). Mater Sci-Medziagotyra 14:230–233

Weber (2015) Leca Insufill—lightweight expanded clay aggregate insulation fill material [Online]. Available: http://www.netweber.co.uk/flooring-systems/products/lightweight-aggregate/lecar-insufill.html. Accessed 16 Dec 2015

Wienerberger (2016) Das Original Gefüllte Ziegel von Wienerberger [Online]. http://service.enev-online.de/bestellen/wzi_101221_gefuellte_ziegel_poroton-p_poroton-mw.pdf. Accessed 14 Jan 2016

Yegian MK, Kadakal U (2004) Foundation isolation for seismic protection using a smooth synthetic liner. J Geotech Geoenviron Eng 130:1121–1130

Yoshihara H, Maruta M (2020) Measurement of the shear properties of extruded polystyrene foam by in-plane shear and asymmetric four-point bending tests. Polymers 12:47

Yoshihara H, Ataka N, Maruta M (2018) Measurement of the Young's modulus and shear modulus of extruded polystyrene foam by the longitudinal and flexural vibration methods. J Cell Plast 54:199–216

YTONG (2021) Product guide: building the future with Ytong autoclaved aerated concrete [Online]. Available: https://www.xella.co.uk/en/docs/Product-Guide-UK-product.pdf. Accessed 28 Nov 2021

Zayas VA, Low SS, Mahin SA (1990) A simple pendulum technique for achieving seismic isolation. Earthq Spectra 6:317–333

Zegowitz A (2010) Cellular glass aggregate serving as thermal insulation and a drainage layer. In: XI International ASHRAE conference, Clearwater, Florida, pp 1–8

Open Access This chapter is licensed under the terms of the Creative Commons Attribution 4.0 International License (http://creativecommons.org/licenses/by/4.0/), which permits use, sharing, adaptation, distribution and reproduction in any medium or format, as long as you give appropriate credit to the original author(s) and the source, provide a link to the Creative Commons license and indicate if changes were made.

The images or other third party material in this chapter are included in the chapter's Creative Commons license, unless indicated otherwise in a credit line to the material. If material is not included in the chapter's Creative Commons license and your intended use is not permitted by statutory regulation or exceeds the permitted use, you will need to obtain permission directly from the copyright holder.

Chapter 4
Evaluation of Critical Structural Assemblies

4.1 Detail Evaluation Methodology for Energy-Efficient Buildings

Based on the literature review and current practice regarding energy-efficient buildings and their details (Chap. 3), we proposed a methodology with which structural details on the building envelope could be evaluated. The methodology was devised on the basis of various regulations (environmental, energy, construction, etc.), and facilitates the evaluation of details in view of environmental and energy-efficiency and structural requirements, which are frequently contradictory. We attempted to illustrate the complexity of designing details in energy-efficient buildings characterised by numerous requirements aiming to ensure that details correspond to technical and structural as well as environmental and energy-efficiency conditions. The structural detail design concept includes various disciplines and experts, particularly architects, and civil and mechanical engineers. When designing structural details, they must provide particularly:

- a clever design taking into account the determined structural assemblies (the material of the load-bearing structure, the position and material of protective layers—waterproofing, thermal insulation, etc.);
- sufficient structural safety for all foreseen load cases;
- the prevention of thermal bridges or good energy efficiency;
- durability and sustainability;
- thermal comfort, etc.

Several impact parameters can be simply incorporated with the proposed methodology, allowing for the recognition of problematic details and facilitating the search for solutions for their improvement. It can be used to design new buildings, or for energy or structural renovation of buildings.

The problem of energy-efficient buildings in earthquake-prone areas, whose crucial element is a well-designed earthquake-resistant structure that complies with the requirements for earthquake-resistant construction, was particularly considered in the devising of the methodology. The proposed methodology can be used to recognise which details are or are not useful in earthquake-prone areas, and whether improvements from the aspect of resistance against seismic action are possible. The exploration of the energy efficiency of details, which is part of the methodology, is an interface for testing various structural solutions when designing the detail. On their basis, measures to improve the energy efficiency of the building envelope can be proposed. The sustainability of the structure and sustainable materials exhibit growing influence on building design, which is why the methodology also includes the life cycle assessment. The monograph applies the principle of environmental evaluation based on indicators for sustainable building evaluation adopted by the CEN TC 350 working group (BMWBS 2019; CEN 2010). It should be noted that the importance of certain aspects and parameters changes in each specific case of a detail. The methodology takes this into account with various weighting factors and external parameters, which cannot be foreseen at the conceptual level or directly included in the overall assessment of each detail.

The development of materials and new technologies significantly influence the design of new details. In many cases, new detail concepts are used to push the boundaries also in the structural sense. This could mean, in certain cases, that no regulations or standard solutions are available to help structural engineers ensure structural safety. Such a case is, for example, the analysis of the detail of foundations on thermal insulation in which regard progress in the development of thermal insulation materials fostered the installation of thermal insulation under the foundation slab. This made thermal insulation the load-bearing layer, which had not been possible and conceivable before. Similar progress is expected in the future, particularly with the development of 'all-round' materials that will meet all the required aspects, significantly contributing to improvements in details. With the proposed methodology, it can be critically estimated which properties of new details deviate significantly from the current practice, and how this affects the technical and structural, and environmental and energy-efficiency parameters of the detail by applying the basic principles of engineering design.

To obtain a suitably supported and objective assessment of structural details, a wide range of quantitative and qualitative criteria must be met. Crucial is an in-depth insight into plans (all floor plans and cross-sections, the building façade area, plans for details, reinforcement plans, etc.) to obtain as much information on the characteristics of materials, geometry, the configuration of a contact, the complexity of construction, the position in the envelope, and other important properties as possible. The practical value of the methodology for evaluating structural details lies on the applicative, and partially, on the theoretical and educational level. The main objectives for it to be used in practice can be summarised as follows:

- a lack of methods for evaluating structural details through the prism of seismic safety;
- the recognition, comparison, and analysis of critical details on the envelope of energy-efficient buildings;

4.1 Detail Evaluation Methodology for Energy-Efficient Buildings

- the review and analysis of the situation in the field of earthquake resistance of details in energy-efficient buildings;
- to raise awareness of the importance of structural details, and their impact on the seismic safety of buildings, the prevention of thermal bridges, and on the provision of thermal comfort for users;
- to facilitate progress in earthquake engineering and architecture by designing good energy-efficient and earthquake-resistant details;
- to foster cooperation between disciplines involved in the design of energy-efficient details, and co-create structural design particularly in the initial phase of the building design;
- to eliminate inappropriate details in the construction of new buildings and renovations; and
- to contribute to better knowledge and recognition of the construction methods of energy-efficient buildings in earthquake-prone areas.

Below, our proposal of environmental and energy-efficiency parameters (Sect. 4.1.1), technical and structural parameters (Sect. 4.1.2), and external parameters (Sect. 4.1.3) for the evaluation of structural details is presented. In Sect. 4.2, all the selected parameters for the environmental and energy-efficiency evaluation are presented in more detail. In a similar way, Sect. 4.3 includes a description of all parameters for the technical and structural evaluation, while Sect. 4.4 covers all external parameters that can affect the total score of a detail. Section 4.5 contains a description of the assessment method, which is based on individual scores for each parameter, and the influence of weighting factors and external parameters on the final score. The assessment method is summarised from (Slak 2010; Slak and Kilar 2008; URBEM 2004), where a similar methodology is used in a different context to assess the characteristics of watercourses and earthquake architecture. The last section of this Chapter covers the limitations of the methodology and potential ways of graphical result presentation.

4.1.1 Importance of Environmental and Energy-Efficiency Parameters

Preventing thermal bridges in the building envelope and the energy aspect of detail design are crucial and cannot be disregarded in modern energy-efficient buildings like in the past. Recently, attention has also increasingly been paid to reducing the environmental impact by using sustainable materials, which can be fully/partially recycled or reused and whose production requires less energy. Such an approach to building design will contribute to reducing greenhouse gas emissions, optimising the use of raw materials, and considering other aspect of sustainable building design. When devising the detail evaluation methodology, we attempted to define evaluation parameters to cover the so-called green evaluation as accurately as possible.

Environmental and energy-efficiency aspects are also recognised in the strategic document (UN 2016; EU Commission 2014) and the recast Constructions Product Regulation (No 305/2011) which adds two requirements: energy economy and heat retention, and sustainable use of natural resources. According to the recast EU regulation, the latter is part of the basic requirements for construction works. The regulation states that the construction works and their heating, cooling, lighting and ventilation installations must be designed and built in such a way that the amount of energy they require in use is low, when the number of the occupants and the climatic conditions of the location are taken into account (construction works must also be energy-efficient, using as little energy as possible during their construction and dismantling). Regarding sustainability, the regulation contains a requirement that the construction works must be designed, built and demolished in such a way that the use of natural resources is sustainable and the following is ensured: (i) the reuse or recyclability of the construction works, their materials and parts after demolition; (ii) the durability of the construction works; and (iii) the use of environmentally compatible raw and secondary materials in the construction works. Additionally, the environmental and energy-efficiency detail evaluation includes the basic requirement of the regulation that throughout their life cycle, the construction works must not be a threat to the hygiene or health and safety of workers (e.g. the giving-off of toxic gas, the emissions of dangerous substances, the emission of dangerous radiation, the release of dangerous substances into ground and drinking water, faulty discharge of waste water, the emissions of flue gases, faulty disposal of solid or liquid waste, dampness in parts of the construction works or on surfaces within the construction works. etc.).

Based on the basic requirements of the Constructions Product Regulation (No 305/2011), the requirements of the passive house standard (Passivhaus Institut 2012), and other applicable legislative documents on efficient use of energy in buildings, and the requirements for modern energy-efficient buildings (see, for example, IBO (2008), Dequaire (2012), John and Zeumer (2015), Desideri and Asdrubali (2018)), the following criteria for the environmental and energy-efficiency detail evaluation were devised: (i) the thermal transmittance of structural assemblies; (ii) the continuity of thermal insulation; (iii) condensation and thermal comfort; (iv) the influence on energy use; (v) airtightness; (vi) the life cycle assessment (LCA); and (vii) durability and stability. The assessment criteria and the characteristics of each parameter are described in Sect. 4.2. For relevant detail assessment, each part of the evaluation must be understood to provide the best estimate in view of the criteria.

4.1.2 Importance of Technical and Structural Parameters

The structural resistance of details is paid an even closer attention than the environmental and energy-efficiency parameters. We decided on such an approach on the basis of the requirement for continuous thermal insulation on the envelope of energy-efficient buildings, which can occasionally affect structural safety. Therefore, possible interventions in the load-bearing structure of individual critical details

4.1 Detail Evaluation Methodology for Energy-Efficient Buildings

were presented in previous chapters due to the requirements for preventing thermal bridges and assumptions on their impact on structural safety are pointed out. Particular emphasis is on how changes in details due to energy efficiency can affect seismic safety. Based on previous studies Azinović et al. (2014a, b, 2015, 2016), it can be concluded that the realisation of good structural details that are crucial to a controlled seismic response of a structure is extremely challenging. For this reason, the parameters directly or indirectly related to structural safety in earthquake-prone areas are taken into account in the second part of the evaluation.

When designing the continuous thermal envelope, the most demanding task is the crossing of two or more structural assemblies. The continuity of the load-bearing structure must be ensured first. Only after that all other protective layers including thermal insulation (Fig. 2.2) could be placed. In certain cases, this results in a thermal bridge in the conceptual design of a detail. The solution should not be the weakening of the load-bearing structural elements on the account of thicker thermal insulation. This is even more complex in the case of earthquake-resistant structures, since changes in some important details could more significantly affect the global response of a building. The locations of considered details in many cases coincide with positions of potential plastic hinges that could be formed during a strong earthquake (such as column-foundation or column-beam connections). The capacity design method which should be used in earthquake prone areas, requires that certain parts of the structure fail before the other parts do and, in such a way, tries to protect the life safety of more important elements. The relations between the dimensions, strengths and ductility of different structural joints should not be changed on the account of thicker thermal insulation, because they might change the desired plastic mechanisms of the structure. Such undesired mechanism could be, for instance, a soft storey, a weak storey or another partially plastic mechanism, which could possess much smaller earthquake response capacity of the structure. The proposed detail evaluation methodology tries to consider these specifics of detail design in earthquake-prone areas by specifying if the considered detail is appropriate in earthquake-prone areas or not.

Well-designed structural details are among crucial parts of a building structure, which can be understood from the basic requirement on mechanical resistance and stability referred to in the Constructions Product Regulation (No 305/2011). This requirement stipulates that the construction works must be designed and built in such a way that the loadings that are liable to act on them during their constructions and use will not lead to (i) the collapse of the whole or part of the work; (ii) major deformations to an inadmissible degree; (iii) damage to other parts of the construction works or to fittings or installed equipment as a result of major deformation of the load-bearing construction; and (iv) damage by an event to an extent disproportionate to the original cause. When devising the detail evaluation methodology, we, in addition to these principles, considered particularly the principles referred to in Eurocode 8 regarding structure design, which is extremely important for earthquake resistance. Eurocode 8 requires in the early stage of structure design to strive for (i) structural simplicity; (ii) continuity and symmetry; (iii) static indeterminacy; (iv) sufficient resistance and stiffness in two horizontal directions; (v) sufficient torsional strength and stiffness;

(vi) a suitable connection between load-bearing elements and interstorey slabs or other structures which act as rigid diaphragms; and (vii) appropriate foundations.

The aforementioned leading principles of Eurocode 8 on the design of load-bearing structures were indirectly or directly incorporated in the proposed parameters for the technical and structural detail evaluation, including (i) the load-bearing capacity; (ii) minimum dimensions and stiffness; (iii) the symmetry of a detail; (iv) the continuity/uniformity of the load-bearing structure; (v) the eccentricity of a detail or a shift in the structure according to the primary load-bearing axis; (vi) the capacity design method; and (vii) connections between primary and secondary (non-)load-bearing elements (e.g. fixing secondary (non-)load-bearing elements). Each parameter is described in more detail in Sect. 4.3. Additionally, the final score also depends on external parameters (Sect. 4.4), and weighting factors for each parameter (Sect. 4.5). The quality of the detail assessment depends mainly on the accuracy and quantity of information collected on the detail. If the quality of concrete or the quantity and orientation of rebars are not known, it is difficult to determine the strength of the building connection detail between two reinforced concrete cross-sections. Assessments of technical and structural parameters can be based on experimental results of the load-bearing capacity, a challenging numerical analysis of a detail using the finite element method, and as a last resort, we can use our experience in comparing several details and examples of poor practice.

4.1.3 Importance of External Parameters

When devising the detail evaluation methodology, we aimed at making it applicable to all the critical details in energy-efficient buildings. Given the wide range of structural details that may occur in every energy-efficient building, their final score in relation to the environmental and energy-efficiency, and technical and structural parameters cannot be expected to be comparable. In addition, details can, on the basis of their basic score, be evaluated wrongly if observed merely locally. Their influence on the global seismic safety of the building structure and the influence on energy use for the whole building must also be taken into account. Therefore, so-called external parameters related to the basic score were defined to include other influences, making scores more comparable and relevant for each specific building.

We defined the following six parameters according to additional external actions that may change the basic technical and structural, and environmental and energy-efficiency assessment: (i) location; (ii) the importance of the building; (iii) the influence on the global analysis; (iv) the complexity of construction; (v) penetrations and openings; (vi) the economic aspect. All parameters are presented in more detail in Sect. 4.4 and the names of parameters are also very revealing. For example, the 'location' parameter refers to the importance of the location of the analysed building. After defining the location, standard climatic conditions can be prescribed and the level of seismic hazard can be determined (seismic hazard function or design ground acceleration). If the location parameters are known, we can decide on more or less complex details. If we are at a location with a high seismic hazard and cold climate, all the parameters of a detail for providing seismic safety and energy efficiency are very important, and can be additionally weighted with an external parameter (increasing or reducing the final score).

4.2 Environmental and Energy-Efficiency Parameters

4.2.1 Thermal Transmittance of Structural Assemblies (E1)

Heat in structural assemblies is transferred mainly or fully with conduction or heat transmission, whereby heat in the winter time (in cold climate regions) flows from the building interior with higher temperatures to the exterior with lower temperatures. The law of heat conduction (known as Fourier's law) applies to such heat conduction. It can be derived from Fourier's law that heat flow resulting from conduction is proportional to temperature gradient, whereby proportionality constant λ (W/m K) called thermal conductivity is used. Based on thermal conductivity, two quantities frequently used in building structures, i.e. thermal resistance R and thermal transmittance U, can be defined. With these quantities, the influence of the thickness of structural assemblies on the thermal protection of a building is taken into account.

In modern energy-efficient building, thermal transmittance must be as low as possible to get a well-insulated building envelope. Therefore, it is among the indispensable parameters in the detail assessment. The criteria of minimum thermal transmittance in various structural assemblies are used in almost all regulations on energy efficiency. Table 4.1 shows thermal transmittance limitations as seen in developed EU member states with a continental climate and PH standard (Passivhaus Institut 2012).

Table 4.1 Prescribed thermal transmittance values (U_{max}) for the building envelope in developed EU countries with continental climate and PH standard

Prescribed thermal transmittance U_{max} (W/(m² K))	Current min. value for EU countries	Standard PH
Outer walls towards unheated rooms	0.28	0.15*
Slab on ground	0.35	0.15
Walls bordering unheated adjacent buildings	0.50	/**
Glazing	1.10	0.80
Window frames	1.30 (1.60)***	0.80

*The PH standard recommends a value for residential buildings with up to two storeys, which stands at 0.10 W/(m² K)
**The PH standard does not provide any value
***1.3 for windows with a timber frame and 1.6 W/(m² K) for windows with a metal frame
Source Summarised from MOP (2010) and Passivhaus Institut (2012)

Limit minimum values are not prescribed and methodology users can select them by themselves in correspondence with the applicable regulations in their country. The detail assessment criteria are determined on the basis of these values. The assessments are primarily narrative but can also be determined more in detail (more in Sect. 4.5). Details with structural assemblies whose U-value is lower than required by the applicable legislation to obtain a building permit would generally get the worst assessment. The 'satisfactory' assessment could be given to details with thermal transmittance lower or equal to the value determined in the applicable rules or technical guidelines, but not reaching PH standard requirements (Table 4.1). Values higher than required by the applicable legislation are inadmissible and categorised as 'poor'. The best assessed are building connection details, where all connected structural assemblies are characterised by a thermal transmittance lower than required by the PH standard.

An additional problem when assessing the thermal transmittance of structural assemblies is the fact that the analysed detail can include several different structural assemblies or other building elements. For example, all connected structural assemblies in the detail may be well insulated and comply with the PH standard, but inappropriate glazing and a frame with a high thermal transmittance are used in windows. In this case the detail cannot get the best assessment, the reduced assessment should be approximately determined by interpolation of scores. In this way also intermediate assessments between poor/satisfactory/good can be obtained.

4.2.2 Continuity of Thermal Insulation (E2)

A low thermal transmittance of all structural assemblies does not guarantee a good thermal response of structural details. It is also important for a good thermal response of a detail whether thermal insulation in the analysed detail is continuous and without interruptions. This parameter is used to recognise details in which thermal bridges occur. At the conceptual level, the locations of potential interruptions of thermal insulation due to structural assembly connections are presented in Chap. 3. Based on the analyses, it can be established whether thermal bridges are unavoidable in certain structural assemblies. If the case that thermal bridges are unavoidable, solutions should be found for such detail or the detail needs to be replaced.

The requirement on the continuity of thermal insulation complies with the basic requirements of the PH standard according to which thermal bridges are undesirable. Interruptions in thermal insulation resulting from the load-bearing structure or other disruptions lead to thermal bridges, which is why such details' assessments are the poorest. Details with interrupted thermal insulation could be rated as "satisfactory" only if at the interruption the material with better insulation properties is used. Attention must be paid to the measurements of the thermal conductivity of material in various directions. Such an example was presented in Sect. 3.2 on base insulation blocks which have good insulation properties perpendicularly on the thermal bridge, making them not equivalent to thermal insulation. Therefore, they can be assessed as

4.2 Environmental and Energy-Efficiency Parameters

'satisfactory' in the best-case scenario. Only details with continuous thermal insulation can receive the best assessment. Details with continuous thermal insulation, which is obstructed in any way, can be assessed between 'satisfactory' and 'good'. Such an example is the insulation of a reinforced concrete parapet on the edge of a flat roof (see Fig. 3.23), where thermal insulation goes around the parapet. From the energy aspect, this is worse than continued insulation with a thermal insulation penetration, as energy losses are higher.

4.2.3 Condensation and Thermal Comfort (E3)

The continuity of thermal insulation significantly affects the occurrence of condensation in structural details and thermal comfort provided to users by the well-insulated building envelope, making the assessment parameters closely related. The interruption of thermal insulation is a condition for low temperatures on the surface of the detail, but it is not a sufficient condition for condensation. In addition to low temperatures, condensation occurs if the relative humidity is high. For this reason, controlled mechanical ventilation devices are installed in modern energy-efficient buildings to improve indoor living conditions and reduce relative humidity. This section includes the criteria for condensation used for detail assessment. They are based on the presumed standard conditions (surface temperatures and relative humidity are determined on the basis of climatic conditions for the analysed building location). The conditions for thermal comfort, which are the most subjective parameter in the process of assessing the environmental and energy-efficiency parameters, are also provided.

If the detail's surface temperature (θ_{si}) drops below a certain limit value, condensation can occur at this location, resulting in mould and lower thermal comfort. Low surface temperatures can also be a reason for the resident's discomfort, as the same thermal comfort requires a higher air temperature (θ_i) at a lower surface temperature (Table 4.2). When assessing thermal comfort, the position of thermal insulation is also important. It is more favourable if it is on the external side of the structural assembly, enabling a linear temperature drop in the assembly.

In certain cases, a lower surface temperature resulting from thermal bridges in combination with a high humidity can lead to condensation, as water vapour is among gases in the air. The amount of vapour in the air depends on air temperature. Vapour condenses onto another surface only if that surface is cooler than the dew

Table 4.2 Pairs of internal air (θ_i) and surface (θ_{si}) temperatures equivalent to the operative temperature of 21.1 °C

θ_{si} (°C)	18.3	19.4	20	20.6	**21.1**	21.7	22.2	22.8	24.4	25.0	25.6	26.1	26.7
θ_i [°C]	25.0	23.4	22.7	21.9	**21.1**	20.3	19.6	18.8	16.4	15.7	14.9	14.1	13.3

The values in bold present a boundary case of equal surface and internal air temperature
Source Krainer (2011)

point temperature defined as temperature at which water vapour condenses at constant pressure. The definition of relative humidity, which is the ratio between the maximum amount of humidity at a certain temperature and the actual amount of humidity, is relevant to the understanding of the condensation process. Condensation occurs in certain part of building elements due to a substantial drop in the surface temperature (θ_{si}) in comparison with the indoor temperature (θ_i), reaching dew point temperature or 100% relative humidity near the surface with a lower temperature.

Condensation can be determined using computer programmes and the hygrothermal analysis of the structural detail (see, for example Delgado et al. (2012)). The suitability of details can be assessed with the enthalpy-entropy diagram or the Mollier diagram (see, for example, Eastop and McConkey (1993)). By analysing details, the dew point temperature can be determined for various boundary conditions. At air temperature $\theta_i = 20\,°C$ and relative humidity of 50%, the dew point temperature is approx. $\theta_{100} = 9.3\,°C$. In addition to complete condensation, mould on building elements can be caused by long-lasting high relative humidity (>80%), which stands at $\theta_{80} = 12.6\,°C$ in the case of the analysed boundary conditions. The influence of details' surface temperatures on the occurrence of mould is frequently calculated in relation to the outside temperature (θ_e) with temperature factor (f_{Rsi}) (CEN 2008):

$$f_{Rsi} = \frac{\theta_{si} - \theta_e}{\theta_i - \theta_e}. \tag{4.11}$$

In the proposed methodology, the analysis of the temperature range must show that the surface temperature on a certain part of a structural detail is lower than the dew point temperature ($\theta_{si} < \theta_{100}$ or $f_{Rsi} < f_{Rsi,100}$) for the 'poor' assessment. Details, in which a thermal bridge will not reduce temperature to the point where mould could occur but are assessed as having a negative impact on the users' well-being, are assessed as satisfactory. Thus, it can be estimated that the temperature in the critical region of the building connection detail is not reduced to temperature θ_{80}, meaning the conditions for the occurrence of mould will probably not be met in the design situation. In this case, mould is only possible if the outside temperature (θ_e) is reduced to extremely low, unexpected values. Details, in which the surface temperature (θ_{si}) and the indoor air temperature (θ_i) are almost the same, are best assessed. Recommendations for the lowest surface temperatures that still ensure thermal comfort are provided in, for example, Parsons (2014). Such details are characterised by the correct distribution of thermal insulation on the structural detail, their thermal bridges are completely eliminated, and materials with suitable heat capacity of the final layer are used to improve thermal comfort.

4.2.4 Influence on the Use of Energy (E4)

Since the calculation of the use of energy in a building is complex, it is difficult to unambiguously determine the value of linear thermal transmittance for a thermal bridge (ψ), which would constitute a good and poor detail respectively. For these reasons, there are few assessments of detail suitability from the aspect of thermal protection. The principles for the influence of details on the use of energy are included in technical guidelines (e.g. MOP (2010)), where the recommended limit coefficient for thermal bridges is $\psi = 0.2$ W/(m K). Before the construction of passive houses emerged, certain recommendations for thermal bridges were prepared as part of the *Eurokobra* project (Janssens et al. 2007), in which the basic decisive assessment criteria for detail suitability can be found (Table 4.3). If the coefficient drops below this conservative estimate, the influence of thermal bridges on the use of energy is deemed negligible and does not have to be taken into account, However, (Janssens et al. 2007) carried out studies of conventional buildings with a lower thermal envelope quality than required for energy-efficient buildings, in which energy consumption is significantly lower. Nevertheless, such assessment criteria are still useful to evaluate the suitability of details in existing building stock. For energy-efficient buildings, the influence of thermal bridges relatively increases, since the

Table 4.3 Limit values for the linear thermal transmittance (ψ_{lim}) of building details with reduced effect on heat loss

Detail typology	Limit linear thermal transmittance coefficients ψ_{lim} (W/(m K))
Building connections on the envelope (junctions at exterior corners): • Roof eaves (façade, gable, etc.) • Façade above overhanging floor	0
Building connections on the envelope (junctions at interior corners): • Roof junction with upper wall • Façade below overhanging floor	0.15
Cantilever structures for balconies	0.10
Details by openings on the envelope: • Window sills, lintels, • A roof window, dormer	0.10
Building connections of the load-bearing structure between: • The roof or an outer wall and inner (non-) load-bearing wall • An outer wall or inner wall and the foundation slab • An outer wall and the interstorey slab	0.05

Source Summarised from Janssens et al. (2007)

remainder of the building envelope is considerably better thermally insulated. Therefore, the requirement to prevent thermal bridges from the PH standard (Passivhaus Institut 2012), according to which thermal transmittance is limited for thermal bridges ($\psi < 0.01$ W/(m K)), is pointed out. The influence of details with thermal bridges on the use of energy can be determined with the calculation of energy balance using the PHPP tool (Passive House Planning Package). For passive house certification, exterior dimensions are used to calculate ψ, however, some building codes are using also internal dimensions.

To determine the detail score according to the use of energy parameter, the thermal analysis of the detail must be performed or losses must be read from the thermal bridge atlas, which provides values for countless details. The thermal analysis of structural details is most frequently carried out with tools based on modelling according to the finite element method. The characteristics of structural assemblies (e.g. the thermal conductivity, thickness and geometry of materials, etc.) and the boundary conditions of the calculation (outside and inside design temperature, and relative humidity) must be taken into account. Details with a thermal transmittance coefficient essentially higher than allowed by the PH standard ($\psi \gg 0.2$ W/(m K)) are assessed as the poorest. Details with coefficients higher than 0.01 W/(m K) are assessed as satisfactory. However, they may still be used in energy-efficient buildings if their impact on the use of energy is not significant (in comparison with the values in Table 4.3). The best assessed details must correspond to the value from the PH standard ($\psi \leq 0.01$ W/(m K)), which could mean in practice a negative value of the coefficient ψ. Such details can be deemed details in which thermal bridges are prevented (their impact on the use of energy is not significant) and do not have to be taken into account in the energy use calculation in the building. A significant weighting factor in the assessment of the impact of structural details on the use of energy is the length of a thermal bridge. It is not taken into account in the basic score, but its impact can be captured by external factors, i.e. the external parameter of impact on the global computational analysis (Sect. 4.4.3) and the parameter on the importance of the detail or building (Sect. 4.4.2). If, for example, we analyse an important detail extending along the whole building perimeter, its impact on the use of energy in the building will be high due to the great length of the thermal bridge and exposure to low temperatures.

4.2.5 Airtightness (E5)

Airtightness denotes the intensity of uncontrolled air flow through the structure into or from the building due to differential pressure (e.g. Zbašnik-Senegačnik (2007), Fennell and Haehnel (2005)). Uncontrolled air flow occurs in gaps, cracks and other leaks on the building envelope. Such locations can lead to a poorer quality of the living environment, which is why they are not allowed in energy-efficient buildings. The latter was mentioned in the previous chapters, as airtightness is discussed as one of the basic concepts of energy-efficient buildings and is also required by the

Fig. 4.1 Investigations of the building envelope airtightness: in search of leakages with a handheld anemometer. *Source* Building and Civil Engineering Institute ZRMK (authors: Praznik, M. and Malovrh, M.)

PH standard. Air flow through the building envelope can result in: uncontrolled heat losses, a negative impact on thermal comfort, building damage, sound transmission, etc. The PH standard determines the maximum value $n_{50} \leq 0.6\ \mathrm{h}^{-1}$, which means that the 50 Pa differential pressure through all leaks in the house extracts 60% of the whole inner air volume in the house in one hour. The blower door test is used to establish leaks in the building, during which air intrusion into sensitive details is measured with sensitive measuring devices under the differential pressure in the building (Fig. 4.1).

Problems that can arise by disregarding airtightness should be avoided in the early stage of design, whereby the design of structural details and the selection of the load-bearing structure are also important. Particularly important is the design of details, as all contacts and penetrations must be envisaged and presented in a detailed plan before construction. Therefore, the detail assessment takes into account the aspect of airtightness intended to prevent leaks in the building envelope and point this aspect out when designing details. The basic principles of the design of an airtight envelope must be considered when assessing details, i.e.: (i) whether the plane of the airtight envelope is uninterrupted in all parts of the building; (ii) there must only be one airtight plane—leaks are not eliminated with an airtight plane; and (iii) an airtight envelope is always attached to the internal side of the thermal envelope (it can also act as a vapour barrier). The airtightness of the materials that constitute the building envelope must be checked first when designing an airtight plane. The airtightness of the conventional materials that constitute the envelope varies significantly (Table 4.4). Materials with $q_{50} < 0.1\ \mathrm{m^3/(m^2\ h)}$ can be generally considered as airtight, whereby q_{50} indicates the air volume permeated through a square meter of the material in one hour at the differential pressure of 50 Pa. Also crucial is the fact that the selected plane is closely connected with adjacent airtight planes. In addition, efficient airtightness of the envelope depends on the connections of individual building elements, in which alternating airtight planes between the internal and external side of the structure must be avoided. The latter means that the contacts between the same structures of building elements (e.g. solid reinforced concrete walls, interstorey slabs and the

Table 4.4 Air permeability at the differential pressure of 50 Pa for various materials

Material	Air permeability q_{50} in m³/(m² h)
Lime-cement plastering (minimum possible thickness)	0.002–0.05
Brick*	0.001–0.05
Autoclaved aerated concrete	0.06–0.35
Plasterboards	0.002–0.03
Plywood	0.004–0.02
Chipboard	0.05–0.22
Solid fibreboards	0.001–0.003
Soft wood fibreboards	2–3.5
Polyethylene foil, 0.1 mm	0.0015
Bitumen cardboard	0.008–0.02
Mineral wool	13–150

*In masonry structures, bricks and grouts as basic materials are not problematic from the aspect of airtightness, but inappropriate gaps in-between different layers
Source Zbašnik-Senegačnik (2007)

roof) are much less complex. In addition to the stated reasons, many leaks in the building envelope are caused by penetrations in structural assemblies with various installations, which cannot be avoided in certain cases.

Details without ensured airtightness and wind tightness are assessed as the poorest (due to visible penetrations in the load-bearing structure, changes in geometry, different layers in the structural connection assemblies—difficult to reach airtightness, inappropriate materials used in the airtight plane (see Table 4.4) Details with difficult contacts of the airtight plane and (or) in which all principles for designing an airtight building envelope can be assessed as satisfactory. The airtightness parameter is closely connected with the external parameters of the complexity of construction (Z4), and penetrations and openings (Z5). If a detail is composed of many different layers and is difficult to build, its score can be further reduced on the basis of parameter Z4. Similarly, all openings and penetrations (Z5) in the envelope significantly affect airtightness.

4.2.6 Life Cycle Assessment (E6)

Life Cycle Assessment (LCA), a methodology defined by international standards ISO 14040/14044 (ISO 2006a, b), addresses the environmental aspects and potential environmental impacts (e.g. resource use and environmental consequences of emissions) throughout the life cycle of products and processes, from raw material extraction to production, use, end-of-life treatment, recycling, and final disposal (i.e. cradle-to-grave) (Röck et al. 2020; Asdrubali and Grazieschi 2020; Klöpffer

4.2 Environmental and Energy-Efficiency Parameters

and Grahl 2014). This technique provides a sound methodological basis for calculating energy demand and assessing resource use, greenhouse gas (GHG) emissions and other environmental indicators throughout the life cycle of buildings and their components (Lasvaux et al. 2016). Methodological developments in recent years have successfully enabled the application of LCA in the construction industry and for buildings (Rasmussen et al. 2018).

There is a wide range of construction solutions and products for buildings that meet different energy-efficient solutions for the design of energy-efficient buildings. LCA can assist architects and other decision makers, i.e. clients, building professionals, and policy makers, in the planning and design of new buildings, and to identify opportunities to improve and optimise the environmental performance of products at different points in their life cycle (Trigaux et al. 2020). Therefore, as an important part of the environmental-energy assessment of any detail in the design of energy-efficient buildings, the LCA data (such as Environmental Product Declarations (EPD) (Passer et al. 2015), eco-labels, building certifications, and other formats that follow ISO 14040/14044) can be used to determine which building materials and construction methods are the most suitable for that detail from the environmental perspective. The data used to calculate the environmental impact indicators and the indicator describing resource use in this book were obtained from the IBO Catalogue of Reference Values for Building Materials. The catalogue was compiled at the end of 2007 for the study "Passive House Building Element Catalog" (IBO 2008), and is constantly updated and extended. The environmental data for the general processes, such as energy systems, transport systems, basic materials, disposal processes and packaging materials, are largely taken from ecoinvent v2.1, while the calculations for the building materials assessment are performed using SimaPro software based on the CML Baseline 2001 method. The assessment is carried out over the whole life cycle (cradle-to-grave, modules A1-5, B1-5, C1-2, C4, D, in accordance with EN 15804 (CEN 2019), which sets the core rules for the preparation of EPDs) during a reference study period of 100 years. For the environmental assessment of each detail, we used the following environmental and resource use indicators: (i) total non-renewable primary energy (PENRT) (ii) global warming potential (GWP) (iii) acidification potential (AP), and (iv) aggregated indicator OI3.

Based on the described environmental evaluation, we can form scores for each detail considered. The worst rated details are those that use only materials with a higher environmental impact. The details rated as satisfactory are those in which materials with a moderate environmental impact are predominantly used. Only components with embedded materials that have a lower environmental impact over the life cycle received the top score., e.g. timber elements (Lukić et al. 2020; Unuk et al. 2021), wood wool and cellulose as thermal insulation (Dickson and Pavía 2021; Casas-Ledón et al. 2020; Hill et al. 2018) etc.

4.2.7 Durability and Sustainability (E7)

All energy-efficient buildings are related to the term 'sustainable development' or 'sustainable construction', while in a general and expert vocabulary, this term covers a wide range of requirements for building design. In both professional and scientific literature (see, for example, Ding (2008), Hajdukiewicz et al. (2015), Halliday (2008), Kibert (2008), Nicol and Humphreys (2002), Hill and Bowen (1997)), the term 'sustainable development' is defined as a balance of four aspects: economic, environmental, social and scientific. The proposed methodology takes into account all the aspects of sustainability directly or indirectly related to the design of structural details. For example, the environmental aspect was taken into account with the life cycle assessment parameter (E6), the potential for the health impact was taken into account with the condensation and thermal comfort (E3) parameter, and the economic aspect was taken into account with the external parameter (Z6). On the other hand, the term 'durability' defined to assess details is used in another context in relation to terms, such as resistance, stability and resilience. For example, durability of timber products is defined with classes of the threat of a biological attack and the classification of the natural durability of wood. Similarly, the durability of steel structures can be defined with the risk of corrosion and in certain details of material fatigue. The durability of reinforced concrete structures can be connected with the thickness of the concrete cover layer, cracking, and classes of exposure to external actions.

The detail is assessed negatively if any of the required protective layers is missing or is interrupted at certain locations. On the other hand, if all protective structures are present, the details could be satisfactory or good, depending on their complexity.

4.3 Technical and Structural Parameters

4.3.1 Load-Bearing Capacity (K1)

The load-bearing capacity is used to assess the resistance of each detail in terms of internal stresses resulting from compressive, tensile, torsion, bending and shear forces. The load-bearing capacity can be established more precisely by experimentally testing the detail. In general, the analysis of the load-bearing capacity of individual materials or structural assemblies does not mean that the load-bearing capacity of the whole detail is known, as there are many parameters that influence the load-bearing capacity of the detail. All structural details are composites or hybrids of several materials and elements. Therefore, all their properties and relations between individual materials and components must be known.

In addition to experimental analyses, detail numerical models and calculation programmes (software) for analysis using the finite element method (FEM) can be used to determine the load-bearing capacity of the detail, whereby all (nonlinear)

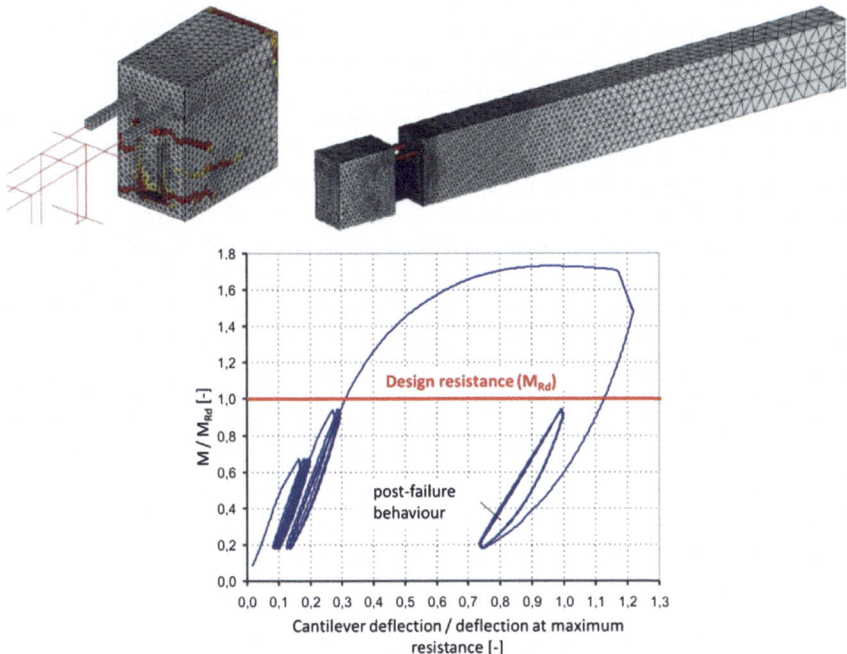

Fig. 4.2 FE-model to investigate the precast load-bearing thermal insulation element (top) and force–deformation curve for specimen with concrete edge failure (bottom). *Source* Heidolf and Eligehausen (2013)

properties of materials or components included in the detail must be known. Figure 4.2 shows an example of the numerical model for the nonlinear analysis of load-bearing thermal insulation elements. The result of a numerical analysis of structural details are usually shown graphically as a force–deformation diagram (Fig. 4.2). In general, determining the load-bearing capacity for a structural »+« type connection detail, for which the load-bearing capacity must be determined in various directions (horizontally, vertically) and for various stress states, which may occur at such a connection, is the most complex.

In addition to the aforementioned options, the load-bearing capacity of a detail can be assessed, in the extreme case, on the basis of experience and the basic engineering judgement and logic. Such an approach is chosen particularly if no relevant data are available on the load-bearing capacity of the analysed detail or such a capacity cannot be directly determined with an experiment (e.g. in the case of the building restoration). The load-bearing capacity of individual components (precast elements, connectors, fasteners, etc.) and materials that make up the detail may be taken into account, and on this basis, the load-bearing capacity of the whole detail can be assumed. To simply determine the load-bearing capacity, the principle of determining such a capacity for the weakest member of the primary load-bearing structure may be taken into account, and on this basis, the load-bearing capacity of the whole detail can be assessed. With the experiential approach, data from the literature on comparable details, whose load-bearing capacity has been precisely tested or thoroughly analysed, may be used.

If the load-bearing structure in the detail is poorly designed (e.g. due to unsuitable connections in steel structures, insufficient reinforcement in reinforced concrete cross-sections, etc.) for vertical static loads and seismic action, the detail's assessment is poor. Details with good load-bearing capacity under vertical static loads, but poor load-bearing capacity for severe seismic action, are assessed as satisfactory. This means that, for example, the principles of earthquake engineering were not considered when designing the detail, resulting in the detail's lack of ductile behaviour under cyclic loading. Based on this fact, it can be established that the detail will be damaged under severe seismic action and should be replaced after the earthquake. Only details with good load-bearing capacity in the ultimate limit state (ULS) under both vertical static loads and dynamic seismic loads get the best score. In the event of severe seismic action, minor or no damage to such a detail is expected.

4.3.2 Minimum Dimensions and Stiffness (K2)

Stiffness is defined as a force generated with the displacement of the structure, or part or element of the structure. Stiffness was also taken into account when assessing structural details in energy-efficient buildings. The stiffness of individual building connection details is determined by the basic dimension of the elements (columns, beams, walls, slabs, etc.), connections between the elements (steel connectors, fasteners in timber structures, the anchorage of reinforcement, etc.), and the load-bearing structure material (the properties of the material—e.g. the elastic modulus in the elastic analysis). Therefore, determining stiffness is similarly complex to determining the load-bearing capacity and must be determined for each analysed detail. Experimental research, the FEM analysis (if available) or any other simplified method can be used for that. On the basis of such analyses, the stiffness of individual structural details can be deducted from the force–deformation diagram, whereby stiffness is represented by the slope of the curve. Simplified approaches and simplifications to determine the stiffness of building structures are frequently used on the basis of geometrical limitations and minimum dimensions of building elements. Therefore, the stiffness parameter is supplemented by the notion of a minimum dimension.

The ability of a detail to deform in the nonlinear range is called ductility, which is closely related to the load-bearing capacity and stiffness of a detail in the nonlinear range. In relation to the level of ductility, Eurocode 8 prescribes the ductility classes—low (DCL), medium (DCM) and high (DCH), and the corresponding complexity of the structural detail design. To design earthquake-resistant structures, Eurocode 8 and other standards stipulate that structures must be designed to be characterised by energy dissipation and to provide energy dissipation capacity and an overall ductile behaviour. For the required overall ductility of the building to be achieved, the potential locations for plastic hinge formation (locations where seismic energy is dissipated) must possess high plastic (rotational) capacities. To this end, ductile failure modes (e.g. flexure) should precede brittle failure modes (e.g. shear) with sufficient reliability. In structures with a high ductility class, structural details are the most complex (e.g. in reinforced concrete structures, smaller spacing is required between stirrups in critical regions, the required steel reinforcement in edge columns of a reinforced concrete wall is higher, etc.).

Eurocode 8 also stipulates certain geometrical limitations related to individual structural elements (e.g. minimum wall dimension) and contacts in the connections of the load-bearing structure (critical region, etc.). The building connection details that observe all code provisions regarding minimum dimensions are best assessed. Directly related to stiffness is the external parameter of influence on the global analysis (Z3), which may be used to reduce the score of the details whose stiffness is crucial to the global seismic response of the whole structure.

4.3.3 Symmetry (K3)

Symmetry or asymmetry of the load-bearing structure can significantly affect the earthquake resistance of a building, particularly due to the diversity of dynamic seismic loads, which act in all directions (see, for example, Koren (2011), Anagnostopoulos et al. (2015), Etedali and Sohrabi (2016), Laguardia et al. (2019), Barbagallo et al. (2020), Tsourekas et al. (2021)). To assess structural details, the symmetry of the detail in relation to the main horizontal and vertical load-bearing axis is determined. The term 'symmetry' includes the symmetry of a detail in relation to the load-bearing structure, connectors, geometry (for example, whether there are penetrations, openings or other structural interruptions on one side of the load-bearing axis), etc. The adverse effects of an asymmetrical detail on structures in earthquake-prone areas may be illustrated with a joint in the steel structure (Fig. 4.3).

Bolts are placed particularly in the upper part of the analysed joint, as it is primarily designed to take on loads resulting from vertical static loads. Such a joint is fully load bearing and rigid under loads with a negative moment resulting from vertical static loads. On the other hand, such a joint is no longer fully load bearing (its load-bearing capacity is lower than the load-bearing capacity of the attached beam (see diagram in Fig. 4.3)) for loads with a positive moment. The direction of the bending moment could change under seismic action, which means a significantly reduced load-bearing capacity for the specific asymmetrical joint, which is lower than planned for static loads. The latter could lead to damage to the load-bearing structure, increased structural deformations, and other undesired effects brought on by seismic action. A similar case of an asymmetrical detail is precast load-bearing thermal insulation elements used in cantilevers. The seismic risk of such elements is considerably higher than of conventional reinforced concrete cantilevers on account of their asymmetrical longitudinal reinforcement (the cantilever uplift can cause damage to the lower part of the cross-section, since the detail lacks bottom steel reinforcement). In addition to the aforementioned, symmetry also greatly affects the ductility of a detail.

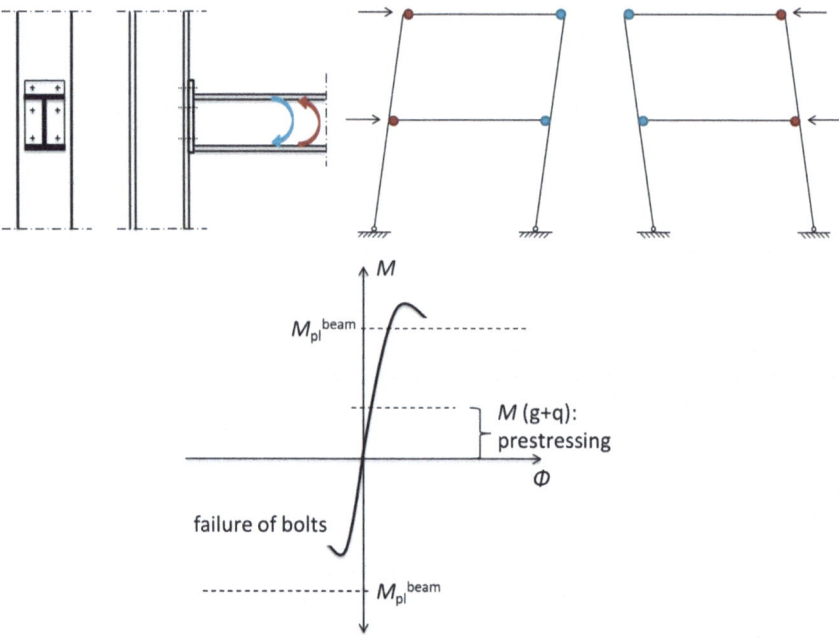

Fig. 4.3 Response of the asymmetrical steel beam-column joint subjected to seismic action

Each structural detail is scored on the basis of a close examination of its geometry and all other building structure plans. Explicitly asymmetrical details are assessed as poor. Symmetry requirements are also referred to in Eurocode 8 (in articles 5.2.3.7.e, 5.5.3.1.3., and 6.5.2.(1)) (CEN 2005). Asymmetrical details are usually taken from non-earthquake-prone areas and designed only for vertical static loads. Details in which the principles of earthquake engineering (e.g. no additional steel reinforcement in the compressive zone of the RC element, the distribution of bolts in a steel structure is not symmetrical, etc.) are violated are assessed more poorly. Details that are not fully symmetrical, but the consequences of asymmetry do not significantly affect the seismic safety of the detail, are assessed as satisfactory. Such effects include asymmetry due to material, which is not necessarily bad in the event of seismic loads. Only details that are completely symmetrical, fully load bearing, and ductile in all directions of expected loads can get the highest score.

4.3.4 Continuity or Uniformity of the Load-Bearing Structure (K4)

The basic principles of designing earthquake-resistant structures include uniformity and continuity of the load-bearing structure. Uniformity in the floor plan is

4.3 Technical and Structural Parameters

characterised by an even distribution of the elements of the load-bearing structure, which allows short and direct transmission of the inertia forces created in the distributed masses of the building. Also important is the uniformity of the structure along the height of the building, which prevents the occurrence of sensitive zones, where concentrations of stresses or large ductility demands might prematurely cause collapse. Continuity is also prescribed in Eurocode 8 and addressed particularly to prevent various irregularities in the plan and along the height of the building. For example, a continued physical connection between horizontal and vertical seismic ties is required in masonry structures. Similarly, continuity requirements also refer to masonry or concrete infills. Irregular, asymmetrical and uneven distribution of infills, and various quantities of openings in infills may lead to floor plan irregularities and irregularities along the height, which can, in certain cases, significantly change the seismic response of the structure. The term 'continuity' is also used to assess structural details, observing changes in the load-bearing structure materials, the number of openings and other possible interruptions.

Details interrupted in the horizontal and vertical directions of the load-bearing structures are assessed as poor. Details with poor connections between structural assemblies at the building connection detail (e.g. due to discontinued connectors, insufficient anchorage length of the steel reinforcement, etc.) are also assessed as poor. In terms of continuity, a detail can be assessed as satisfactory if the load-bearing structure is interrupted, for example, by a change in the material, but well connected to each other by appropriately designed connectors or other binding elements. Continuity may be satisfactory in precast elements, in which the load-bearing part of the structure is well connected by connector elements and whose load-bearing capacity is able to withstand all seismic action. Details, in which there are no interruptions due to changes in the material, insufficient connector elements or large openings, are the best assessed. If different materials are connected, they must act uniformly as a whole, and the properties of all the materials used in the detail do not differ significantly, making them useful in earthquake-prone areas (see, for example, articles 5.2.3.4 c, 5.3.2 and 6.2 in Eurocode 8).

4.3.5 Eccentricity or a Shift in the Structure According to the Primary Load-Bearing Axis (K5)

In earthquake engineering, the term 'eccentricity' is usually used in relation to the distribution of masses, the load-bearing capacity, and/or stiffness. When assessing structural details, the term 'eccentricity' is used in relation to the displacement of the load-bearing structure in relation to its primary load-bearing axis. This principle is directly and indirectly observed by seismic codes which, for example prescribe that the eccentricity of the beam axis relative to that of the column into which it is fixed must be limited to enable an efficient transfer of cyclic moments from a primary seismic beam to a column to be achieved. Examples of poor practice include the discontinuity of the load-bearing structure resulting from the requirements for energy efficiency and continuous thermal insulation.

The geometrical properties of a detail referred to in the plan for the load-bearing structure of a building can be used to score the eccentricity parameter. Details, whose load-bearing structure in the building connection detail is eccentrically displaced in relation to the load-bearing axis in the structural system of a building, get the poorest score. It could be estimated that the unacceptable eccentricity of the building connection detail is significant if the load-bearing structure is displaced by over $1 \cdot d$, whereby 'd' is the thickness of the structural assembly. Such details can be used to increase the thickness of thermal insulation, penetrations for installations, and other energy efficiency measures. Details, in which displacing the load-bearing structure does not pose a special threat to the transmission of forces to other structural elements and the transmission of forces is ensured through other load-bearing elements that eliminate the adverse effects of eccentricity, are assessed as satisfactory. This means that details with small displacements of the load-bearing structure can be assessed in such a way. Details without eccentricity in the horizontal and vertical directions (all load-bearing elements are placed relative to the main load-bearing axis in the structural system) get the best score. The transmission of forces in the building connection detail is not hindered in any direction, and all the requirements referred to in Eurocode 8 or similar standards for seismic resistant structures are met (e.g. regarding regularity in elevation, geometrical limitations for beams, etc.).

4.3.6 Capacity Design Method (K6)

The capacity design method is a design method, in which the selected elements of the structural system are designed and structured to dissipate energy at great deformations. All other elements are provided with the load-bearing capacity that supports the selected manner of energy dissipation. This method is used to determine the hierarchy of the load-bearing capacity of various elements of load-bearing structure to provide a suitable plastic mechanism and prevent brittle failure modes. If dissipative zones are located in the structural elements, the non-dissipative parts and the connections of the dissipative parts to the rest of the structure must have sufficient overstrength to allow the development of cyclic yielding in the dissipative parts. The method principle can be explained by imagining the structure as a chain with a ductile weak link (Fig. 4.4), as proposed by (Paulay and Priestly 1992).

The force in the chain F_E cannot exceed the value determined with the actual load-bearing capacity of the weak link $R_{act,du}$:

$$F_E \leq R_{act,du} \qquad (4.12)$$

If the load-bearing capacity at the yield point of all stronger links $R_{CD,nd}$ is greater than the actual load-bearing capacity of the weak link $R_{act,du}$ (*CD* index marks capacity design):

4.3 Technical and Structural Parameters

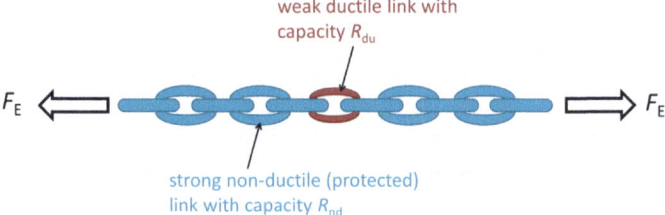

Fig. 4.4 Analogy between the structure and a chain with a ductile weak link. *Source* Summarised from Fajfar et al. (2008), *Original source* Paulay and Priestly (1992, p. 40)

$$R_{CD,nd} > R_{act,du}, \quad (4.13)$$

the inelastic deformations (damage) will be limited to the weak link acting as a seismic fuse. However, ductility must be provided to the weak link.

Structural details are given the poorest assessment if they are not designed according to the earthquake engineering principles, which may result in the formation of a plastic hinge in the wrong structural element. Details, which are not protected against brittle failure are assessed as poor; for example, foundations or columns and other important sections of the vertical structural system which are not protected. Details, which do not support the capacity design method but whose influence of the global earthquake resistance is negligible, are assessed as satisfactory, which means that this is a building connection of the load-bearing structure, which will not result in a global failure, the instability of the structure, etc. Details, which were designed for earthquake-prone areas, provide global ductility and plastic hinges can form at the planned locations in the structure. All principles referred to in Eurocode 8 (e.g. Article 6.2.3, etc.) should be observed.

4.3.7 Connections of Primary and Secondary (Non-) Load-Bearing Elements (K7)

The technical and structural evaluation of a detail can be significantly affected also by secondary (non)load-bearing elements, e.g. fixing façade panels or protective layers for the outer wall structural assembly. The anchorage of façade panels in energy-efficient buildings is complicated in certain cases, as great thickness of thermal insulation makes it difficult to ensure a full-strength connection with the primary load-bearing structure. In addition, strong connections between the secondary and primary load-bearing structure in energy-efficient buildings are not desired, as most such connections result in thermal bridges. For this reason, new precast elements for fixing on the envelope, which combine load-bearing and thermal insulation properties, are made, but are not tested to be used in earthquake-prone areas. In addition to façade panels, other secondary (non-)load-bearing elements, which are part of the building envelope (e.g. windows, balcony doors, roof windows, fences, gutters, etc.) are problematic. All these elements in the envelope constitute significant costs of energy-efficient buildings and their failure or severe damage could even put the lives of users of the building at risk. In the methodology, the effect of secondary (non-)load-bearing elements on the structure or seismic safety is evaluated according to the suitability of the connection with the primary load-bearing structure to which these elements are fixed.

To provide a relevant assessment of the suitability of connections between the primary load-bearing structure and secondary (non-)load-bearing elements, the fixing details, the capacity of connectors, the ductility of the secondary substructure, and other similar properties must be known. Details are assessed more poorly if the connection between secondary elements and the primary load-bearing structure is interrupted at a crucial location to provide resistance to vertical static loads and seismic action. Deductions are also possible if the fixing of secondary elements in the critical regions of an earthquake-resistant structure is planned. In reinforced concrete structures, for example, Eurocode 8 defines critical regions as regions, where the most adverse combination of action effects (M, N, V, T) occurs and where plastic hinges may form. If secondary elements are anchored or fixed in the critical regions of the load-bearing structure, this could mean additional internal forces in this part of the load-bearing structure, which must be considered when designing. Details without a critical fixing of secondary elements, and with the primary load-bearing structure and no anchorage or penetrations in the critical regions of the load-bearing structure, where safety could be reduced in cases of vertical static loads and seismic action, get the highest score. All articles of Eurocode 8 or similar design standards for earthquake resistant structures must be taken into account to ensure the ductile behaviour of connections or structural details (e.g. Article 5.1.2 in Eurocode 8).

4.4 External Parameters

External parameters are used to correct the basic score for the technical and structural, and environmental and energy-efficiency evaluation. In the proposed methodology, external parameters may hold values between 0 and 2, and their description consists largely of an assessment in terms of positive and negative impacts on the basic scores. If the influence of an external parameter is significant, the basic score is increased (use a factor higher than 1) and reduced if the influence of the external parameter is low (a factor lower than 1). By default, the external parameters have the value of 1, meaning they have no influence on the evaluation.

4.4.1 Location (Z1)

In environmental and energy-efficiency parameters, location affects the outside temperature and humidity, resulting in different input project data for energy-efficient buildings. In technical and structural aspects, the influence of a parameter may be even greater if the area is seismically very active. The climate data of environmental agencies in EU member states or similar global data may be used to assess the complexity of boundary climatic conditions. Similarly, seismic hazard of a location can be seen on a seismic hazard map, which includes design ground acceleration to determine design seismic actions according to the deterministic design approach. If the probabilistic approach is used, the seismic hazard function must be used to determine the seismic hazard. Based on these data, the assessor may give an external parameter a low value (lower than 1), whereby the influence of the location is low—e.g. a warm climate (high design temperature) and a low seismic hazard of the location (e.g. design ground acceleration $a_g < 0.10$ g). The highest value of the influence of the location is captured by factor 2, which reduced the influence of the basic parameters. This factor must be used to multiply the basic parameters with explicit influence of the location. This means that the area is seismically very active with design ground acceleration $a_g > 0.35$ g. In environmental and energy-efficiency parameters, the basic score is reduced in the case of extremely cold areas with a very low design temperature.

4.4.2 Importance of a Building (Z2)

Basic scores may also be corrected on the basis of the importance of a building or part of a building, in which the analysed structural detail being assessed is located. In the first phase, the importance of a building is related to its intended use. On this basis, actions on the load-bearing structure (e.g. imposed load) as well as boundary conditions for living quarters in energy-efficient buildings (indoor temperature, humidity in a room) may be determined. The importance of a building is defined according to input project data in accordance with the applicable standards (e.g. Eurocode 8) and other guidelines (classification according to the guidelines of cultural heritage institutions, cultural, urban, architectural or substantive significance of the building, etc.).

In Eurocode 8, the importance of a building is defined in Sect. 2.1 and takes into account differentiation in relation to the required reliability of a building. An importance factor is directly and indirectly expressed also in non-structural elements, which are part of almost every structural detail. If the importance factor of a building is high, the responsibility in planning details is high. Such buildings include hospitals, which must be fully operational immediately after an earthquake, resulting in stricter requirements for individual details. In such a case, the analysed external parameter may be higher than 1 and the basic technical and structural assessments may be suitable corrected. On the other hand, a lower importance factor means less responsibility for well-designed details, which is why the basic technical and structural scores may be attributed lower importance (Z2 < 1). Similarly, the importance of environmental and energy-efficiency parameters can be determined on the basis of the importance of a building. The Z2 factor of public facilities can be higher than 1.

4.4.3 Influence on the Global Analysis (Z3)

The basis for a relevant and objective assessment from the technical and structural, and environmental and energy-efficiency aspects is good knowledge of a structural detail. The basic scores are particularly part of a detail analysis on the local level (the temperature range of a structural detail is analysed, the load-bearing capacity of cross-sections and connectors are checked for a specific detail, etc.). In theory, this means that a poor assessment of a detail does not signify a poor assessment of the global seismic safety of a building or a significant impact on the use of energy of the whole building. A negative impact on the global structural or seismic safety occurs only when a detail is installed in a critical or vital region of the structure or negatively affects adjacent elements of the structure, worsening their seismic response. By analogy, a structural detail greatly affects the use of energy when the length of the thermal bridge is significant, making it impossible to ignore its global impact. In addition to damage inflicted locally, such a structural detail would also aggravate the global seismic response of a building and significantly affect the use of energy.

A parameter to describe the influence of a basic criterion on the global analysis of a building may be determined on the basis of such a definition. External parameter Z3 may be lower than 1 when a detail in no way affects the global seismic safety. For the structural evaluation, this means that the detail is not part of the primary load-bearing structure to withstand seismic action, enabling higher basic scores. On the other hand, the influence of a detail on the global seismic analysis is recorded if critical details, such as details with which a building is fixed to the foundation, the connection between a column and the unheated basement, connections between the roof and interstorey slabs, etc., are involved. In cases when mere damage to a detail may disrupt static equilibrium, this external parameter can be used to increase influence and reduce the basic scores.

By analogy, negative influence on the global analysis is taken into account in the environmental and energy-efficiency evaluation if a structural detail with high linear thermal transmittance and (or) a lengthy linear thermal bridge is analysed. The use of energy for heating and cooling (energy balance) in energy-efficient buildings must be calculated with precise methods, e.g. the PHPP programme package. Energy-efficient buildings require more complex and sophisticated energy simulations due to lower heat losses to produce most relevant assessment of the influence of thermal bridges. If the calculation with the PHPP method shows significant changes in the use of energy in view of the used structural details, thermal bridges will be considerable and the basic environmental and energy-efficiency scores must be reduced (Z3 lower than 1). Alternatively, if the energy balance according to PHPP shows that the detail is of high quality and has a negligible thermal bridge, a lesser influence on the global response according to the given basic environmental and energy-efficiency scores may be taken into account.

4.4.4 Complexity of Construction (Z4)

The complexity of construction is defined as an external parameter, since correct construction is crucial to a good response of a structural detail throughout its lifetime. Construction errors may result in the detail's greater influence on the use of energy and deterioration in environmental and energy-efficiency parameters. Consequences of poor construction are also possible from the technical and structural aspect (e.g. poor detailing of steel reinforcement on site, poorer materials than prescribed in the project are used etc.). The external parameter of the complexity of construction may be lower than 1 when a standard detail, which is well known to contractors and has been used in practice for years, is involved. The effects of this external parameter may be positive in this case, as it may be confirmed with more certainty that the detail will be both structurally safe and functional from the energy aspect. A requirement from the PH standard is also the acquisition of a certificate for contractors and building products (Passivhaus Institut 2012). The complexity of construction will have a lesser influence also when only verified and certified contractors take part in the design of energy-efficient buildings, and only certified details (windows, structural assemblies, etc.) are installed. Determining greater influence of the complexity of construction takes into account details that are not tested, meaning that solutions contractors do not know are involved, increasing the risk of errors.

4.4.5 Penetrations and Openings (Z5)

Penetrations (e.g. for building installations) in the load-bearing structure of a building may produce great concentrations of stress and hamper the detailing of the load-bearing structure (e.g. the placement and anchorage of steel reinforcement). We find that it is even more important to prevent penetrations or correctly place penetrations in the load-bearing structure in earthquake-prone areas, where changes in the stiffness and load-bearing capacity of structural elements are more undesired. Also, the ductile response of structural details, in which seismic energy dissipation is anticipated, is crucial, and may worsen due to penetrations in the critical regions of the structure. Very important is the influence of all penetrations and openings on the environmental and energy-efficiency parameters, since these are usually locations in the building envelope with high thermal transmittance. In addition, the detail's airtightness must be ensured at the location of penetrations. Since penetrations are possible in almost every structural detail, their influence is covered by an external parameter. In practice, penetrations may be added subsequently and are not foreseen in the load-bearing structure plans. In such cases, a change in the external parameters could be used to check the influence of penetrations on the basic technical and structural, and environmental and energy-efficiency scores of the detail.

From the aspect of structural detail scoring, details without penetrations, whose load-bearing structure is continuous and regular (no negative changes in the load-bearing capacity and stiffness, no changes in thermal transmittance and reduced airtightness, etc.), may get a bonus to the basic score. A lesser influence of penetrations on the technical and structural evaluation may be expected when penetrations are not located in the critical regions of the vertical and horizontal load-bearing structure (e.g. where a beam is fixed to a column). In addition, all the participating designers must foresee in the design phase the locations of penetrations, where stress is not concentrated and the values of internal forces are not increased. On the other hand, the detail's basic score may be reduced if it is interrupted at a crucial location of the load-bearing structure. A more considerable influence of penetrations is recognised also for structural details, in which penetration dimensions are greater than the dimensions of the load-bearing structure. A negative influence of environmental and energy-efficiency parameters may also be expected, as larger penetrations mean a greater influence on the use of energy, and reduced airtightness and other parameters, which cannot be disregarded in energy-efficient buildings.

4.4.6 Economic Aspect (Z6)

Detail solutions can be evaluated also from the aspect of economic justification. This includes the economic justification of the price of the structure and details in earthquake-resistant construction, and the justification of the price of an environmental and energy-efficient solution (the rationality of material choice, structural details with a low energy consumption, etc.). The minimum that would guarantee statutory quality in terms of structural safety and minimum energy consumption must be defined as part of this criterion. Savings related to the earthquake resistance of a structure, which would violate the statutory minimum and provisions on earthquake-resistant construction (e.g. Eurocode 8), must not be allowed in any case. The added value of quantifying the earthquake resistance of a building according to a probabilistic approach may be used to more precisely determine the economic aspect (Žižmond and Dolšek 2019; Sinković and Dolšek 2020).

The evaluation of economy when taking into account the environmental and energy-efficiency principles of a structural details is also very complex (Noureldin and Kim 2021; Shen et al. 2021; Noshadravan et al. 2017). On the one hand, reduced energy consumption for heating and cooling (prevented thermal bridges, greater thickness of thermal insulation, quality glazing, etc.) positively affects monthly costs of heating and cooling, while on the other hand, an investment in the construction phase (restoration, new construction) for such details is higher. Additionally, the European Union and other countries allocated various subsidies to promote energy-efficient solutions, reducing the costs of the initial investment. The study of the economics of a detail is about the greatest effect with the lowest input, while also the desires of the market, users and investors must be taken into account.

4.5 Assessment Based on Evaluation Parameters, Weighting Factors and External Parameters

In the first step of assessment according to the proposed methodology, the assessor or expert assesses the value of individual parameters (E1-E7 and K1–K7) with scores defined in previous chapters. Scores are expressed in numbers, from 0—poor to 6—good. Since different criteria may carry various importance, the methodology support scores (O_{Ei} and O_{Ki}) could be suitably weighted in the next step. The assessor determines the importance or influence of a parameter by determining a suitable weight or weighting factor (γ_{Ei} in γ_{Ki}) for each parameter, which increases or decreases the influence of the scores. The criteria may be weighted evenly (the value of all basic parameters is 1) or unevenly (the basic parameters are weighted and standardised differently).

Decisions on the distribution of weighting factors may be presented in the form of a diagram (Fig. 4.5). Unevenly distributed weighting factors indicate how high the assessor evaluated the importance of a certain parameter, contributing to clarity and more objective evaluation. The ratios between weighting factors are not limited or fixed according to the type of structural details. They may be determined on the basis of the assessment of a specific situation. Nevertheless, it may be established on the basis of a weighted definition of the basic parameter that extreme values of weighting factors are not sensible. Therefore, they are limited in this methodology, with 2 being

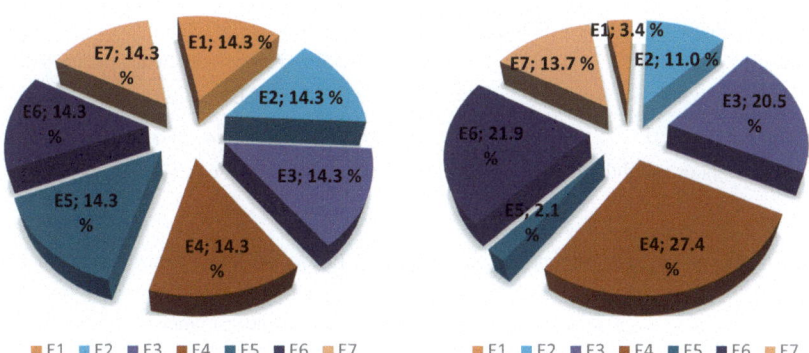

Fig. 4.5 Example of uniformly (left) and unevenly (right) distributed weighting factors for the basic parameters (right) of the environmental and energy-efficiency evaluation of structural details

the highest value of the criteria weighting factors (a high weighting factor is only sensible in exceptional cases when requirements for certain criteria are distinctive). In theory, weighting factors have no lower limit, but the so-called zero weighting factor ($\gamma_{Ei}, \gamma_{Ki} = 0$) can be used to exclude a criterion from evaluation (negative values are not sensible, as the zero factor eliminates the parameter), which is foreseen for special cases, when the evaluation of a certain parameter is not possible or is irrelevant. This means that most important factors receive the weight or importance factor of 2, while less important factors receive, for example, 0.25 (exceptionally 0—the criterion is excluded from evaluation).

In the proposed evaluation methodology, weighting factors must be standardised to produce a comparable sum or total score, which is equivalent to even distribution. The basic scores remain the same, while the level of influence of a parameter on the total score from the environmental and energy-efficiency, and technical and structural evaluation changes with weighting. Weighting factors for individual parameters are used to determine their significance, affecting the final score. A well assessed parameter with a low weighting factor contributes to the total score, while a poorly assessed parameter contributes more due to a high weighting factor, together producing a significantly lower total score. Also important when determining scores and weighting factors is that the environmental and energy-efficiency, and technical and structural parameters are not combined in the total average score, since they can only affect the ratios within one (environmental and energy-efficiency) or another (technical and structural in our case) part of evaluation. Each environmental and energy-efficiency (E1–E7), and technical and structural (K1–K7) parameter is attributed importance (the distribution of weighting factors) with weighting factor γ_{Ei} or γ_{Ki}, and fill in two separate tables (Tables 4.5 and 4.6).

The procedure to determine weighting factors can be schematically shown on three levels. On the first level, the weighting factor of each parameter is defined regardless of the structural detail and the specific use of the building. Determining

Table 4.5 Values of basic parameters and weighting factors for the environmental and energy-efficiency evaluation of structural details

	Environmental and energy-efficiency parameters						
	E1	E2	E3	E4	E5	E6	E7
Achieved score (O_{Ei})	O_{E1}	O_{E2}	O_{E3}	O_{E4}	O_{E5}	O_{E6}	O_{E7}
Selected weight or weighting factor (γ_{Ei})	γ_{E1}	γ_{E2}	γ_{E3}	γ_{E4}	γ_{E5}	γ_{E6}	γ_{E7}

Table 4.6 Values of basic parameters and weighting factors for the technical and structural evaluation of structural details

	Technical and structural parameters						
	K1	K2	K3	K4	K5	K6	K7
Achieved score (O_{Ki})	O_{K1}	O_{K2}	O_{K3}	O_{K4}	O_{K5}	O_{K6}	O_{K7}
Selected weight or weighting factor (γ_{Ki})	γ_{K1}	γ_{K2}	γ_{K3}	γ_{K4}	γ_{K5}	γ_{K6}	γ_{K7}

4.5 Assessment Based on Evaluation Parameters ...

weighting factors on this level depends on the priorities of the designer who devises weighting factors in cooperation with the investor, future users of the building, and the construction technologist. Thus, weighting factors may reflect, on the first level, the input project data, and desires and priorities of the investor or future users of the building. In the addressed technology, the selected parameters on the first level are equal or equally important. Therefore, all the selected weighting factors are still even ($\gamma_{Ei}, \gamma_{Ki} = 1$).

On the second level, general weighting factors must be adapted to the position and significance of each detail in the building envelope. Details in the area of foundations have a different role in relation to the whole building than building connection details between the roof and an outer wall or cantilever structure details for balconies. Differences occur in the assessment of environmental and energy-efficiency, and technological and structural parameters alike. Given the position in the building envelope, the following groups or types of details could be defined: (1) building connections between cantilever elements and the primary load-bearing structure; (2) foundation detail; (3) the contact between the roof and a wall; (4) building connections between interstorey slabs and outer/inner wall; and (5) building connections between a transparent section of the envelope (e.g. windows) and the primary load-bearing structure, etc. The same distribution of weighting factors for the basic parameters of the technical and structural evaluation can be assumed for each group on the second level. However, the type of structural detail connections (»L«, »T« and »+« connection types) must be distinguished within each group and the role of the detail in the thermal envelope must be determined. On this basis, weighting factors for the environmental and energy-efficiency parameters must be defined. Only by determining the role of the detail in the technical and structural, and environmental and energy-efficiency terms, the distribution of weighting factors for parameters may be defined on the second level. The best basis to determine weighting factors on this level is the comprehensive structural and energy consideration of individual groups or types of details to assess the parameters with weighting factors that are more important for certain types of details. Based on the comprehensive analysis of each detail, the monograph provides guidelines to determine weighting factors in the selected cases of details.

However, the assessment does not conclude by determining weighting factors on the second level, as the potential influence of external parameters (third level of determining weighting factors) must also be taken into account. The last (third) level of determining weighting factors is important when assessing the influence of the building local conditions, which is captured in the methodology with external parameters (Chapter 0). The methodology is devised so that external parameters do not directly affect the score, but correct weighting factors. When a weighting factor (γ_{Ei} and γ_{Ki}) is determined, external parameters (γ_{Zi}) increase or reduce it with the selected factor (Tables 4.7 and 4.8). The proposed six external factors may affect each basic parameter differently. Therefore, they are determined separately for the environmental and energy-efficiency (Table 4.7), and technical and structural evaluation (Table 4.8). Like with weighting factors, external parameters are limited to numbers between 0 and 2, whereby the neutral value is 1 and does not affect the

Table 4.7 External parameters influencing the environmental and energy-efficiency evaluation of structural details

	Environmental and energy-efficiency parameters						
External parameters (Z)	E1	E2	E3	E4	E5	E6	E7
Z1–Z6	$\gamma_{Z,E1}$	$\gamma_{Z,E2}$	$\gamma_{Z,E3}$	$\gamma_{Z,E4}$	$\gamma_{Z,E5}$	$\gamma_{Z,E6}$	$\gamma_{Z,E7}$

Table 4.8 External parameters influencing the technical and structural evaluation of structural details

	Technical and structural parameters						
External parameters (Z)	K1	K2	K3	K4	K5	K6	K7
Z1–Z6	$\gamma_{Z,K1}$	$\gamma_{Z,K2}$	$\gamma_{Z,K3}$	$\gamma_{Z,K4}$	$\gamma_{Z,K5}$	$\gamma_{Z,K6}$	$\gamma_{Z,K7}$

assessment. If necessary, the assessor may determine external weighting factors for each basic parameter of evaluation. In general, more or less external factors could be determined, depending on the requirements of the structural detail assessment study.

Tables 4.9 and 4.10 include the influences of external parameters in moderate Central European region according to the previously described guidelines. The values of the criteria included in the tables are guidance values to evaluate details in Central Europe, and cannot be randomly used in evaluations without an analysis and justification.

The total score, which represents the environmental and energy-efficiency, and technical and structural aspects of a detail, is achieved by taking into account the basic scores according to the evaluation criteria, determining weighting factors for individual parameters, and considering the influence of external factors. Expressions

Table 4.9 Influence of external parameters for the environmental and energy-efficiency evaluation of structural details

Weighting factor	External parameters for the environmental and energy-efficiency evaluation					
$\gamma_{Z,Ei}$	Z1	Z2	Z3	Z4	Z5	Z6
0.00	18–20 °C	Energy efficiency class: G	Not important	–	–	Investment is not justified
0.25	16–18 °C	Energy efficiency class: F	$\Delta H'_T < 2\%$	Not complex	Few/no penetrations	
1.00	10–12 °C	Energy efficiency class: C	$\Delta H'_T = 5–10\%$	Medium complexity	Conventional penetrations	Lifetime
2.00	2–4 °C	Energy efficiency class: A1	$\Delta H'_T > 20\%$	High complexity	Many penetrations	In a half of the lifetime

4.5 Assessment Based on Evaluation Parameters ...

Table 4.10 Influence of external parameters for the technical and structural evaluation of structural details

Weighting factor	External parameters for the technical and structural evaluation					
$\gamma_{Z,Ki}$	Z1	Z2	Z3	Z4	Z5	Z6
0.00	Up to 0.10 g	Temporary structures	Not relevant	–	–	Reduced by 100%
0.25	$a_g \cdot S$: 0.10–0.15 g	Importance factor: I	Low importance	Not complex	Few/no penetrations	
1.00	$a_g \cdot S$: 0.20–0.25 g	Importance factor: II	Medium importance	Medium complexity	Conventional penetrations	Basic by regulations
2.00	$a_g \cdot S$: above 0.35 g	Importance factor: IV	Very important	High complexity	Many penetrations	Increased by 100%

used to achieve final values (average scores are calculated by taking into account weighting and external factors) are provided in Eqs. (4.14)–(4.18).

For each evaluation parameter, we first evaluate an average of all external factors:

$$\gamma_{Z,Ei} = \sum_{j=1}^{n_Z} Z_{j,Ei} \cdot \frac{1}{n_Z} \quad \gamma_{Z,Ki} = \sum_{j=1}^{n_Z} Z_{j,Ki} \cdot \frac{1}{n_Z}, \tag{4.14}$$

whereby $\gamma_{Z,Ei}$ is the external factor (Zi) for a certain environmental and energy-efficiency parameter (Ei), $\gamma_{Z,Ki}$ is the external factor (Zi) for a certain technical and structural parameter (Ki), n_Z is the number of external factors, $Z_{j,Ei}$ is the assessed influence of an external factor on a certain parameter (Ei), and $Z_{j,Ki}$ is the assessed influence of an external factor on a certain parameter (Ki).

Corrected weighting factors or weights (Γ_{Ei} and Γ_{Ki}) are determined by multiplying the external factors ($\gamma_{Z,Ei}$ and $\gamma_{Z,Ki}$) and previously determined weighting factors (γ_{Ei} and γ_{Ki}):

$$\Gamma_{Ei} = \gamma_{Z,Ei} \cdot \gamma_{Ei} \quad \Gamma_{Ki} = \gamma_{Z,Ki} \cdot \gamma_{Ki}, \tag{4.15}$$

and are normalised to values between 0 and 1:

$$\overline{\Gamma_{Ei}} = \frac{\Gamma_{Ei}}{\sum_{i=1}^{n_E} \Gamma_{Ei}} \quad \overline{\Gamma_{Ki}} = \frac{\Gamma_{Ki}}{\sum_{i=1}^{n_K} \Gamma_{Ki}}, \tag{4.16}$$

whereby $\overline{\Gamma_{Ei}}$ is a normalised and corrected weight of a certain parameter (Ei), $\overline{\Gamma_{Ki}}$ is a normalised and corrected weight of a certain parameter (Ki), n_E is the number of the environmental and energy-efficiency evaluation parameters, and n_K is the number of the technical and structural evaluation parameters (Ki).

Based on these parameters, the weighted score of parameters (O_{Ei-U} and O_{Ki-U}) is determined, which is expressed as a share of the total score:

$$O_{Ei-U} = O_{Ei} \cdot \overline{\Gamma_{Ei}} \quad O_{Ki-U} = O_{Ki} \cdot \overline{\Gamma_{Ki}}. \tag{4.17}$$

The total weighted average is finally calculated as the sum of all shares of weighted parameters:

$$O_E = \sum_{i=1}^{n_E} O_{Ai-U} \quad O_K = \sum_{i=1}^{n_K} O_{Ki-U}. \tag{4.18}$$

The total weighted average may also be translated into percentage, whereby 100% (or the total final score of 6) signifies an ideal structural detail or the most quality solution (energy-efficient and earthquake-resistant structural detail).

Chapter 5 provides several examples, in which the proposed methodology for the assessment of structural details is used. Concrete examples enable us to imagine the aforementioned equations in a general form even more clearly. Neutral weights and neutral external factors are most frequently adopted. Such an approach was chosen for a better comparability of the assessments of individual details and a relatively low influence of external factors on the total score. More information on the influence of the weighting factors is provided in (Slak 2010; Slak and Kilar 2008) and in two concrete examples (Sects. 5.1 and 5.2). For better clarity, the results are presented in the form of a radial diagram and column chart (Sect. 4.6).

4.6 Presentation of the Results

The evaluation of structural details can be graphically presented in the form of a radial diagram or a floral diagram of criteria, based on which the value of the structural detail from the environmental and energy-efficiency aspect as well as its technical and structural resistance or the level of seismic safety can be shown. Figures 4.6 and 4.7 show two hypothetical examples of evaluation in radial diagrams. The environmental and energy-efficiency parameters are on the left (in green) and the technical and structural parameters are on the right (in red). Each diagram also includes average values numerically presenting the assessment for each evaluation separately (the descriptive assessments of poor, neutral and good with intermediate values). A standard deviation is also provided to show whether individual scores deviate significantly from the average, which may be used as additional information in the detail evaluation. The analysis of the results of the structural detail evaluation may show the interaction between both halves of the floral diagram or the harmonisation between the consideration of the environmental and energy-efficiency, and technical and structural parameters. The influence of weighting factors and external parameters provided with a standardised corrected weighting factor is expressed as percentage next to the diagram and the width of the section of a circle. More information on the influence of weighting factors and external parameters is provided in Sects. 5.1 and 5.2.

4.6 Presentation of the Results

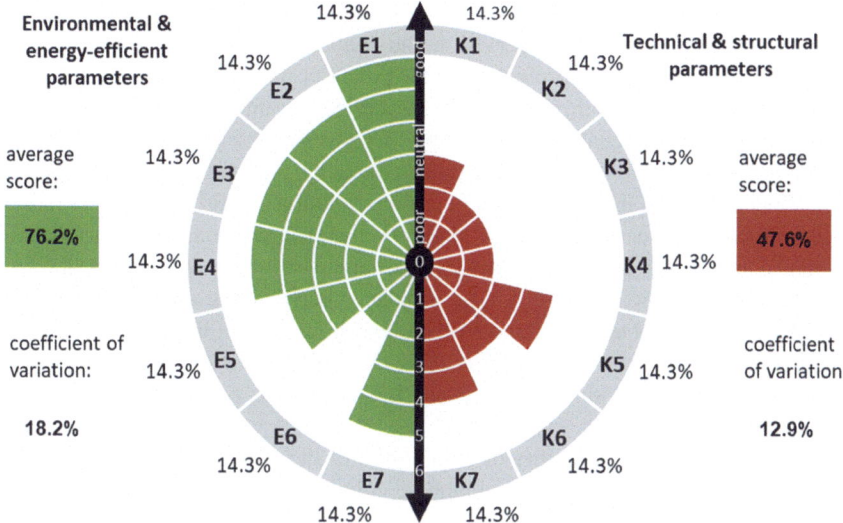

Fig. 4.6 Example of results for the proposed methodology presented in a radial diagram for uniformly distributed weighting factors: environmental and energy-efficiency evaluation (left), and technical and structural evaluation (right)

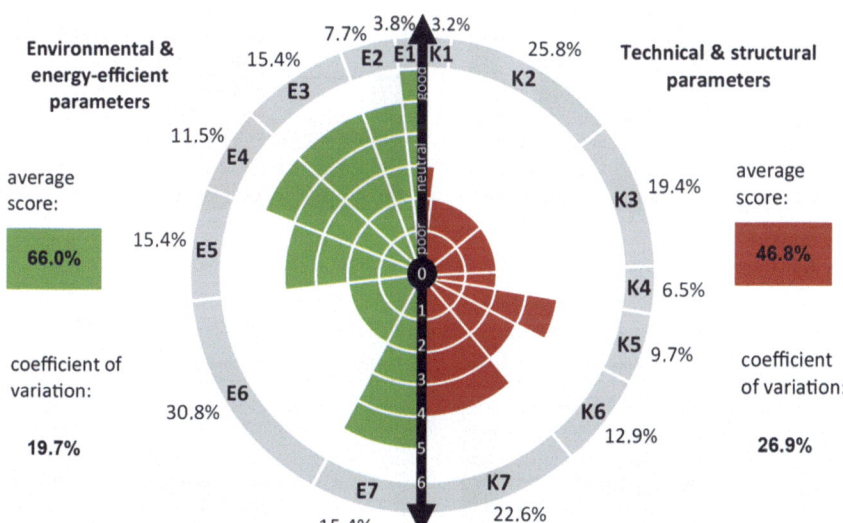

Fig. 4.7 Example of results for the proposed methodology presented in a radial diagram for unevenly distributed weighting factors: environmental and energy-efficiency evaluation (left), and technical and structural evaluation (right)

Figure 4.7 shows a radial diagram with uneven distribution of weighting factors, facilitating better assessment supervision and enabling us to swiftly recognise the aspects of the detail to be improved. Note the potential subjectivity of scores, which may be relatively high if the assessor is inexperienced, unqualified or unsuitable. A radial diagram also enables us to simply detect if any of the parameters in the evaluation system is too overlooked or favoured. Such a system fosters easier assessment supervision and the comparison of various solutions. A fuller form of a radial diagram means higher quality of the structural detail in the context of earthquake resistance and energy efficiency. On the other hand, a distinctly asymmetrical form of a radial diagram would signify a poor balance between the environmental and energy-efficiency, and technical and structural requirements of the structural detail. The superiority or inferiority of one part of the evaluation and inappropriate solutions of structural details can also be recognised if a diagram is partially or completely empty.

To highlight the comparability of both halves of the evaluation, the results may also be shown in the form of a comparative column chart (Figs. 4.8 and 4.9). Its

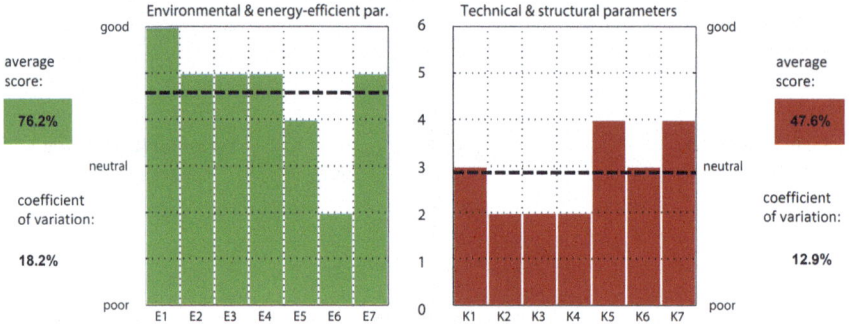

Fig. 4.8 Results of a hypothetical detail evaluation with uniformly distributed weighting factors presented in a column chart

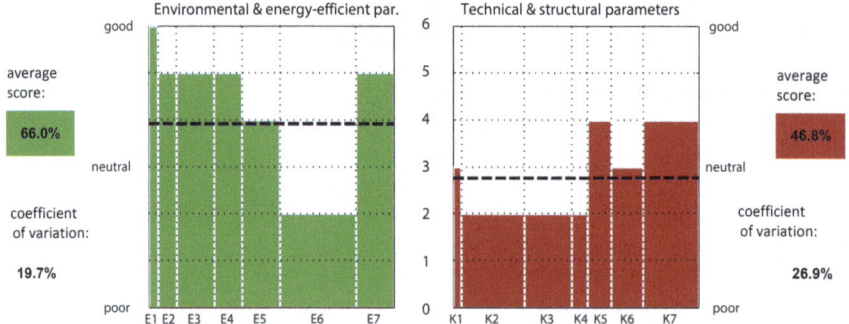

Fig. 4.9 Results of a hypothetical detail evaluation with unevenly distributed weighting factors presented in a column chart

4.6 Presentation of the Results

concept is similar to that of a radial diagram and includes similar graphic elements, which explicitly show the achieved scores at first glance. Similar to a radial diagram, the width of a column may be used to show the importance of weighting factor of a parameter (Fig. 4.9). The advantage of such a display is that the achieved score can be determined simply by calculating the area of the diagram on both sides, whereby a larger area means a higher score and a better quality structural detail. The diagram also includes average values shown in black dashed lines. If the average score for the environmental and energy-efficiency, and technical and structural evaluation deviates significantly, it may be detected with a deviation of the dashed line, which points to an unbalanced detail response.

The proposed assessment system and the manner the results are displayed were devised to reflect the duality of the structural detail evaluation (environmental and energy-efficiency, and technical and structural parameters), emphasising the importance of both aspects for the final quality of the detail used in energy-efficient buildings. As presented below (in Chap. 5), the optimum solution may be selected based on the methodology, which suffices for energy efficiency and earthquake resistance. In practice, certain requirements from the environmental and energy-efficiency evaluation are frequently in contrast with the technical and structural requirements. Discrepancies may be detected with the proposed methodology and eliminated in the design and selection phase of a structural detail. The selection procedure of the best structural detail iterative, since improvements in the technical and structural sense can result in worse environmental and energy-efficiency indicators and vice versa.

Figure 4.10 shows the proposed evaluation criterion for the usefulness of details defined on the basis of the final average score shown in a radial diagram. If a detail

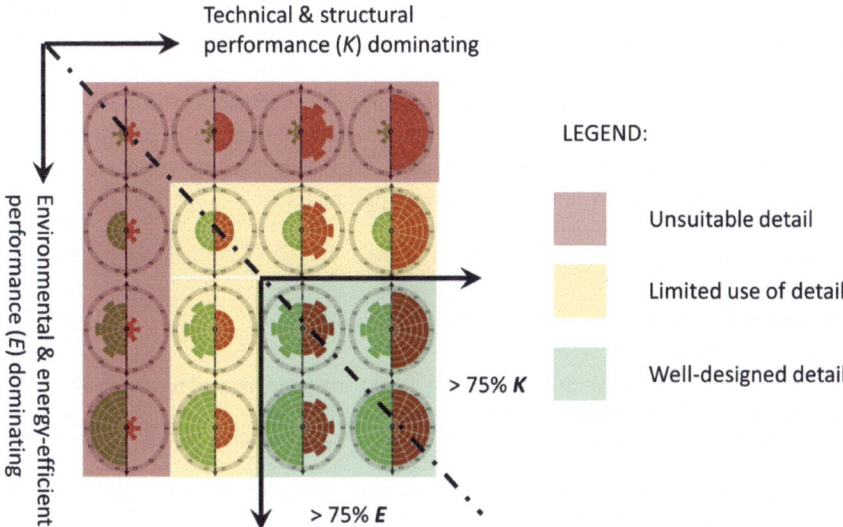

Fig. 4.10 Proposed criteria of structural detail evaluation using a radial diagram for results presentation

receives a total average score lower than 50% from both evaluations, it is treated as unsuitable and must be replaced (red). The latter would mean that the detail has not received a satisfactory score in any of the analysed assessment parameters or certain parameters have been seriously disregarded. If a detail receives a total score of 50% from at least one part of evaluation, one of the requirements for the usefulness of the detail in the specific use of the building is met. Additionally, designers and assessors must assess the influence of weighting factors shown in radial or column diagrams. This also takes into account the ratios between individual parameters, and whether it is sensible to consider the principles of earthquake engineering and energy efficiency.

A well-designed detail is a detail whose total score from both parts of evaluation is higher than approximately 75%. In this case, a good response may be expected from the detail and the consideration of the environmental and energy-efficiency, and earthquake engineering parameters. Based on the proposed methodology, the most suitable structural details may be selected for energy-efficient buildings, which are efficient from the aspect of environmental protection and the use of energy, and do not aggravate the seismic safety of a building. Both selected limit values (50 and 75%) are reference values, and can be sensibly increased or decreased in specific cases. At this point, a safeguard should be added, which cannot be covered by average scores: if the score from a part of evaluation is highly unbalanced (a high standard deviation), the detail is not useful despite its high total average score. The latter is justified by the fact that the use of structural details, in which one of the criteria has been totally disregarded (scoring 0), is not sensible. On the other hand, details with the basic score of 0 can still be useful if the criterion for the detail assessment is deemed irrelevant in the specific case. In such cases, irrelevant criteria may be excluded from evaluation by using the zero weighting factor. This means that, in certain borderline cases, details with 0 basic score are useful if it can be proven that the assessment criteria are not relevant. Such an example is criterion K6 (Capacity design method), which is prescribed in the earthquake-resistant construction standards (e.g. Eurocode 8) and used exclusively for earthquake-resistant structures. In such a case, the methodology is devised to exclude criterion K6 from the evaluation system by using the zero weighting factor if a detail in a seismically less active area is assessed (in seismically inactive areas, this is an irrelevant assessment parameter).

References

Anagnostopoulos S, Kyrkos M, Stathopoulos K (2015) Earthquake induced torsion in buildings: critical review and state of the art. Earthquakes Struct 8:305–377

Asdrubali F, Grazieschi G (2020) Life cycle assessment of energy efficient buildings. Energy Rep 6:270–285

Azinović B, Koren D, Kilar V (2014a) Seismic safety of low-energy buildings founded on a thermal insulation layer. In: A joint event of the 15th european conference on earthquake engineering & 34 general assembly of the European seismological commission, 24–29 August 2014, Istanbul, Turkey

References

Azinović B, Koren D, Kilar V (2014b) The seismic response of low-energy buildings founded on a thermal insulation layer—a parametric study. Eng Struct 81:398–411

Azinović B, Kilar V, Koren D (2015) Erdbebensicherheit vorgefertigter wärmegedämmter Stahlbeton-Konsolenelemente. Bauingenieur 90:489–499

Azinović B, Kilar V, Koren D (2016) Energy-efficient solution for the foundation of passive houses in earthquake-prone regions. Eng Struct 112:133–145

Barbagallo F, Bosco M, Ghersi A, Marino E, Rossi P (2020) A database for assisted assessment of torsional response of in-plan irregular buildings. In: Seismic behaviour and design of irregular and complex civil structures III. Springer

BMWBS (2019) Leitfaden Nachhaltiges Bauen–Zukunftsfähiges Planen, Bauen und Betreiben von Gebäuden

Casas-Ledón Y, Salgado KD, Cea J, Arteaga-Pérez LE, Fuentealba C (2020) Life cycle assessment of innovative insulation panels based on eucalyptus bark fibers. J Clean Prod 249:119356

CEN (2005) Eurocode 8: Design of structures for earthquake resistance - Part 1 : General rules, seismic actions and rules for buildings. EN 1998:2005

CEN (2008) Thermal bridges in building construction—heat flows and surface temperatures—detailed calculations. EN ISO 10211:2008

CEN (2010) Sustainability of construction works—sustainability assessment of buildings—Part 1: general framework. EN 15643–1:2010

CEN (2019) EN 15804: 2012+ A2: 2019: Sustainability of construction works—environmental product declarations—core rules for the product category of construction products

Delgado JM, Barreira E, Ramos NM, De Freitas VP (2012) Hygrothermal numerical simulation tools applied to building physics

Dequaire X (2012) Passivhaus as a low-energy building standard: contribution to a typology. Energy Effi 5:377–391

Desideri U, Asdrubali F (2018) Handbook of energy efficiency in buildings: a life cycle approach. Butterworth-Heinemann

Dickson T, Pavía S (2021) Energy performance, environmental impact and cost of a range of insulation materials. Renew Sustain Energy Rev 140:110752

Ding GKC (2008) Sustainable construction—the role of environmental assessment tools. J Environ Manag 86:451–464

Eastop TD, Mcconkey A (1993) Applied thermodynamics for engineering technologists, Longman

Etedali S, Sohrabi MR (2016) A proposed approach to mitigate the torsional amplifications of asymmetric base-isolated buildings during earthquakes. KSCE J Civ Eng 20:768–776

EU Commission (2014) Horizon 2020 energy efficiency (Online). European Commission. Available: https://ec.europa.eu/easme/en/section/horizon-2020-energy-efficiency

Fajfar P, Fischinger M, Beg D, Dolšek M, Isaković T, Kreslin M, Rozman M, Vidrih Z (2008) Design of building structures according to Eurocode standards, Chapter 8 Design of earthquake resistant structures (in Slovene)

Fennell HC, Haehnel J (2005) Setting airtightness standards. ASHRAE J 47:26

Hajdukiewicz M, Byrne D, Keane MM, Goggins J (2015) Real-time monitoring framework to investigate the environmental and structural performance of buildings. Build Environ 86:1–16

Halliday S (2008) Sustainable construction. Routledge

Heidolf T, Eligehausen R (2013) Design concept for load bearing thermal insulation elements with compression shear bearings. Beton- und Stahlbetonbau 108:179–187

Hill RC, Bowen PA (1997) Sustainable construction: principles and a framework for attainment. Constr Manag Econ 15:223–239

Hill C, Norton A, Dibdiakova J (2018) A comparison of the environmental impacts of different categories of insulation materials. Energy Build 162:12–20

IBO (2008) Details for passive houses: a catalogue of ecologically rated constructions. In: A catalogue of ecologically rated constructions, 3rd edn. Springer, Vienna

ISO (2006a) ISO 14040. Environmental management: life cycle assessment; requirements and guidelines. ISO Geneva, Switzerland

ISO (2006b) ISO 14044. Environmental management: life cycle assessment; principles and framework

Janssens A, Van Londersele E, Vandermarcke B, Roels S, Standaert P, Wouters P (2007) Development of limits for the linear thermal transmittance of thermal bridges in buildings. In: X. International ASHRAE conference, Clearwater, Florida

John V, Zeumer M (2015) Sustainable construction techniques: from structural design to interior fit-out: assessing and improving the environmental impact of buildings, detail

Kibert CJ (2008) Sustainable construction: green building design and delivery. Wiley

Klöpffer W, Grahl B (2014) Life cycle assessment (LCA): a guide to best practice. Wiley

Koren D (2011) Earthquake resistant insulation and irregular layouts in architecture (in Slovene). Doctoral dissertation, University of Ljubljana, Faculty of Architecture

Krainer A (2011) System. In: Module 1, concept of bioclimatic design (in Slovene). Lectures on the subject of Buildings, energy, environment. University of Ljubljana, Faculty of Civil and Geodetic Engineering, Chair of Buildings and Constructional Complexes (KSKE)

Laguardia R, Morrone C, Faggella M, Gigliotti R (2019) A simplified method to predict torsional effects on asymmetric seismic isolated buildings under bi-directional earthquake components. Bull Earthq Eng 17:6331–6356

Lasvaux S, Achim F, Garat P, Peuportier B, Chevalier J, Habert G (2016) Correlations in life cycle impact assessment methods (LCIA) and indicators for construction materials: what matters? Ecol Ind 67:174–182

Lukić I, Premrov M, Leskovar VŽ, Passer A (2020) Assessment of the environmental impact of timber and its potential to mitigate embodied GHG emissions. IOP Conf Ser: Earth Environ Sci 588:022068

MOP (2010) Technical guideline for construction TSG-1-004: 2010, energy efficiency (in Slovene) [Online]. Slovenian Ministry of the Environment and Spatial Planning. Available: http://www.arhiv.mop.gov.si/fileadmin/mop.gov.si/pageuploads/zakonodaja/prostor/graditev/TSG-01-004_2010.pdf. Accessed 22 July 2018

Nicol JF, Humphreys MA (2002) Adaptive thermal comfort and sustainable thermal standards for buildings. Energy Build 34:563–572

Noshadravan A, Miller TR, Gregory JG (2017) A lifecycle cost analysis of residential buildings including natural hazard risk. J Constr Eng Manag 143:04017017

Noureldin M, Kim J (2021) Parameterized seismic life-cycle cost evaluation method for building structures. Struct Infrastruct Eng 17:425–439

Parsons K (2014) Human thermal environments: the effects of hot, moderate, and cold environments on human health, comfort, and performance. CRC Press

Passer A, Lasvaux S, Allacker K, de Lathauwer D, Spirinckx C, Wittstock B, Kellenberger D, Gschösser F, Wall J, Wallbaum H (2015) Environmental product declarations entering the building sector: critical reflections based on 5 to 10 years experience in different European countries. Int J Life Cycle Assess 20:1199–1212

Passivhaus Institut (2012) Certification criteria for residential Passive House buildings [Online]. Darmstadt, Germany. Available: http://www.passiv.de/downloads/03_certfication_criteria_residential_en.pdf. Accessed 18 Aug 2013

Paulay T, Priestly MJN (1992) Seismic design of reinforced concrete and masonry buildings. Wiley, New York

Rasmussen FN, Malmqvist T, Moncaster A, Wiberg AH, Birgisdóttir H (2018) Analysing methodological choices in calculations of embodied energy and GHG emissions from buildings. Energy Build 158:1487–1498

Röck M, Saade MRM, Balouktsi M, Rasmussen FN, Birgisdottir H, Frischknecht R, Habert G, Lützkendorf T, Passer A (2020) Embodied GHG emissions of buildings—the hidden challenge for effective climate change mitigation. Appl Energy 258:114107

Shen Z, Zhou H, Shrestha S (2021) LCC-based framework for building envelope and structure co-design considering energy efficiency and natural hazard performance. J Build Eng 35:102061

References

Sinković NL, Dolšek M (2020) Fatality risk and its application to the seismic performance assessment of a building. Eng Struct 205:110108

Slak T (2010) Characteristics, evaluation and potentials of seismic architecture (in Slovene). Doctoral dissertation, University of Ljubljana, Faculty of Architecture

Slak T, Kilar V (2008) Simplified ranking system for recognition and evaluation of earthquake architecture. In: The 14th world conference on earthquake engineering, China

Unuk Ž, Lukić I, Leskovar VŽ, Premrov M (2021) Renovation of timber floors with structural glass: Structural and environmental performance. J Build Eng 38:102149

Trigaux D, Allacker K, Debacker W (2020) Environmental benchmarks for buildings: a critical literature review. Int J Life Cycle Assess 1–21

Tsourekas A, Athanatopoulou A, Kostinakis K (2021) Maximum mean square response and critical orientation under bi-directional seismic excitation. Eng Struct 233:111881

UN (2016) Transforming our world: the 2030 agenda for sustainable development

URBEM (2004) Urban river basin enhancement methods, work package WP2–WP11. Project Report

Zbašnik-Senegačnik M (2007) Passive house, Ljubljana, University of Ljubljana, Faculty of Architecture

Žižmond J, Dolšek M (2019) Formulation of risk-targeted seismic action for the force-based seismic design of structures. Earthquake Eng Struct Dynam 48:1406–1428

Open Access This chapter is licensed under the terms of the Creative Commons Attribution 4.0 International License (http://creativecommons.org/licenses/by/4.0/), which permits use, sharing, adaptation, distribution and reproduction in any medium or format, as long as you give appropriate credit to the original author(s) and the source, provide a link to the Creative Commons license and indicate if changes were made.

The images or other third party material in this chapter are included in the chapter's Creative Commons license, unless indicated otherwise in a credit line to the material. If material is not included in the chapter's Creative Commons license and your intended use is not permitted by statutory regulation or exceeds the permitted use, you will need to obtain permission directly from the copyright holder.

Chapter 5
Case Study: Using Methodology to Assess the Selected Details

Based on the theoretical framework of the proposed methodology (Chap. 4), this chapter includes the presentation of the evaluation method with concrete examples of structural details. To select structural details to be assessed, we relied mainly on the solutions for energy-efficient buildings presented in previous chapters and catalogues of structural details for passive houses (see, for example, IBO (2008)). By selecting various structural details, we attempted to include a wide range of solutions. However, it must be noted that scores apply only to the analysed case, since the slightest change in the detail results in a different final score. Over twenty details intended to present the use of the evaluation methodology are presented. For a more general assessment of the suitability of a certain energy-efficient detail to be used in earthquake-prone areas, several solutions should be evaluated and assessments for as many details as possible should be substantiated with detailed experimental and (or) numerical analyses.

Table 5.1 Input project data and requirements for the environmental and energy-efficiency evaluation of the analysed structural details

Outdoor temperature θ_e (°C)	Indoor temperature θ_i (°C)	Relative humidity (%)	Temp. factor $f_{Rsi,\,min}$ to prevent mould
−10	20	50	0.75

The following details were evaluated with the proposed methodology: (i) the building connection between an outer wall and the foundation slab (Sect. 5.1); (ii) the building connection between the load-bearing balcony structure and an outer wall (Sect. 5.2); (iii) the building connection between an outer wall and the unheated basement (Sect. 5.3); and the building connection between an outer wall and the roof structure (Sect. 5.4). The temperature range and heat flow through the structure were determined for the selected details with *Thermal Bridge Simulation Tool* in the *Archicad* software environment (Graphisoft 2015). On this basis, the factor of linear thermal transmittance (ψ) and the temperature factor (f_{Rsi}) were determined for each detail to define the parameters of the energy-efficient evaluation. The environmental assessment of a detail was based on the environmental and energy-efficiency parameters if materials used in the detail. The table with the environmental parameters of materials is summarised from (IBO 2008). The basis for the technical and structural evaluation is the analyses of the selected structural details and the review of literature on structural details of energy-efficient buildings. The scores of certain details were determined with the experiential approach and are not based directly on the numerical proof of the load-bearing capacity and other characteristics of the technical and structural evaluation.

To determine the temperature range and heat transfer through structural details, uniform boundary conditions determined for central European city with a continental climate (e.g. Ljubljana, Slovenia) (Table 5.1) were used in all cases. These boundary conditions are required as input data for the numerical analysis performed with *Archicad* (Graphisoft 2015). The programme numerically solves the differential equation of heat transfer through structural assemblies and the ground. Details must be defined as three-dimensional (3D) elements of a building. Subsequently, the section or analysed region in which heat transfer is to be evaluated must be determined. After determining the boundary conditions and materials, the location of (un)conditioned spaces and the course of the envelope must be defined in the programme. Then the finite element network is determined in relation to the desired accuracy of the calculation. A denser network brings more accurate results of the calculation (heat flow and the temperature range of the detail will be determined in several discrete points) but prolongs the time required for the calculation. The details of the numerical calculation (the number of iterations, the relative equilibrium, the asymmetrical index, etc.) are shown in (Blocon 2015).

In addition to the input project data for the analysis (Table 5.1), structural assemblies (the selected materials, the thickness of all layers, thermal transmittance (U), etc.) and other characteristics (the characteristics of the elements that prevent thermal bridges, the characteristics of the fixing elements, connectors, etc.) are precisely described for each detail. Other input data for each analysed structural detail are provided graphically with a 2D cross-section of the detail. The simulation results of heat transfer through structural details are shown in diagrams for the temperature range and heat flow (see, for example, Fig. 5.2). The temperature range diagram shows the temperature range for each discrete point in the finite element network, which depends on the selected accuracy of the analysis. The latter is used to determine the lowest surface temperature ($\theta_{si,\min}$) required to assess a detail from the aspect of thermal comfort and to monitor the occurrence of condensation and mould. To monitor the occurrence of condensation, the temperature factor f_{Rsi} (Eq. 3.11) is additionally calculated. The key result of each evaluation is a table with scores for the basic criteria, the selected weighting and external factors, and the final score (radial and column diagrams).

5.1 Building Connection Detail Between an Outer Wall and the Foundation Slab

The selected evaluation example is composed of a reinforced concrete (RC) outer wall and the reinforced concrete foundation slab (Fig. 5.2). The RC foundation slab runs below the entire floor plan of the building, and is thermally insulated with 2×12 cm thick XPS boards and the nominal compressive strength of 400 kPa. Thermal insulation boards can consist of other materials (e.g. EPS, cellular glass, etc.) with better or poorer properties (a lower or higher compressive strength, a better or poorer thermal conductivity). However, a hypothetical detail with characteristics described below is selected for evaluation. The geometrical thermal bridge at the contact between an outer wall and the foundation slab is prevented with additional thermal insulation (Fig. 5.1, detail 1*). The length of additional underground thermal insulation is 100 cm from the external point, while its thickness amounts to 20 cm. Other characteristics of the selected structural details are stated in Table 5.2.

To assess and define external parameters, the analysed structural detail is assumed to be installed in a four-storey residential building with the floor area ratio of 2:1 ($A/B = 16/8$ m) and located in central Europe (i.e. Ljubljana, Slovenia). Based on the location, climatic conditions and the level of seismic hazard can be defined. It is assumed that the building sits on a good foundation base (type A ground) with the design ground acceleration of 0.25 g. The design climatic conditions for the environmental and energy-efficiency evaluation of the analysed detail are stated in Table 5.1. Given the importance factor, the building falls into Category II (regular building) according to the definition in Eurocode 8. Additional penetrations in the area of the analysed connection detail are not planned. A comparison of the evaluation results with and without taking into account external factors is shown for the analysed detail.

A good thermal response of a detail may be substantiated with the heat transfer simulation results presented in Fig. 5.2. The temperature profile for certain boundary

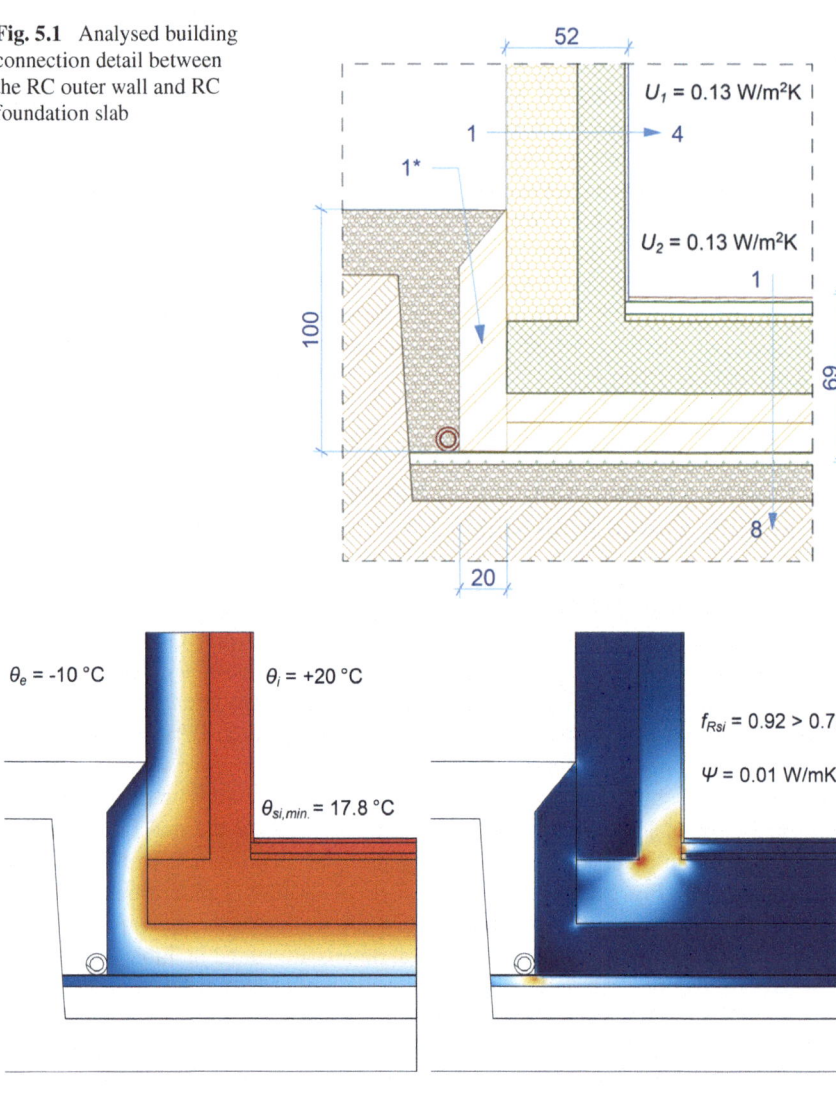

Fig. 5.1 Analysed building connection detail between the RC outer wall and RC foundation slab

Fig. 5.2 Results of the numerical simulation of heat transfer through the building connection detail with a thermally insulated foundation slab

5.1 Building Connection Detail Between an Outer ... 111

Table 5.2 Composition of structural assemblies for the building connection detail between an outer wall and the foundation slab

RC outer wall Thermal transmittance $U_1 = 0.13$ W/(m² K)			Slab on ground—insulated foundation slab Thermal transmittance $U_2 = 0.13$ W/(m² K)		
No.	Material	T (cm)	No.	Material	T (cm)
1	Silicate thin-layer plastering	–	1	Flooring—wooden parquet	1.5
2	EPS (expanded polystyrene)	30	2	Concrete screed	5
3	Reinforced concrete outer wall	20	3	Acoustic mineral wool	3
4	Finishes (plastering, etc.)	–	4	Reinforced concrete foundation slab	30
			5	XPS boards (2 × 12 cm)	24
			6	Bitumen cardboard as waterproofing in two layers	1
			7	Blinding concrete	5
			8	Drainage layer	15

climatic conditions is shown on the left and the pertaining heat flow on the right. We can see from the results that the inner surface temperatures (θ_{si}) are very high. The minimum surface temperature is reached in the corner of the connection between an outer wall and the foundation slab, standing at $\theta_{si,\ min} = 17.8$ °C. This was to be expected, since in theory, this region contains a geometrical thermal bridge, which is also reflected in the course of heat flow (see the right side of Fig. 5.2). The results on heat flow also show that thermal insulation successfully halts a stronger flow through the soil and prevents a thermal bridge in the critical region. In addition to the temperature range, the analysis may be used to calculate relative linear thermal transmittance (ψ [W/m K]), which is generally used to describe the extent of the thermal bridge and effect on the use of energy. The linear thermal transmittance factor for the analysed structural detail is $\psi = 0.01$ W/(m K). Based on the analysis boundary conditions, the temperature factor was also calculated, amounting to $f_{Rsi} = 0.92$.

According to the environmental and energy-efficiency evaluation, the detail was mainly well assessed (Table 5.3). To determine parameter E1, both structural assemblies in the building connections were deemed to have a lower thermal transmittance than required by the PH standard ($U_{max} = 0.15$ W/(m² K)). Nevertheless, one score lower than the highest score (6) was deducted for the detail, as the PH standard recommends even lower values of thermal transmittance for houses (the value of 0.10 W/(m² K) is recommended). Parameter E2 received the highest score, since thermal insulation is continuous, and the thickness of thermal insulation on the outer wall is increased at the critical spot where a geometrical thermal bridge may occur. Consequently, the detail's surface temperatures are high, which was taken

Table 5.3 Values of basic parameters and weighting factors for the analysed structural detail

Environmental & energy-efficiency parameters	E1	E2	E3	E4	E5	E6	E7
Achieved score (O_{Ei}, O_{Ki})	5	6	5	6	6	2	4
Selected weighting factors (γ_{Ei}, γ_{Ki})	1.50	0.75	2	1.50	0.50	0.25	2
Technical & structural parameters	K1	K2	K3	K4	K5	K6	K7
Achieved score (O_{Ei}, O_{Ki})	4	4	4	2	5	3	4
Selected weighting factors (γ_{Ei}, γ_{Ki})	1.50	2	0.50	1.75	0.25	1.25	0.25

into account in the assessment of thermal comfort (E3). Nevertheless, a minimum deduction from the score was required, as materials with a high thermal capacity (see Table 5.2, installed materials) were installed as finishes, which evokes a cold feeling when touching an outer wall or the floor. Parameter E4 received the highest score, as the details shows a very low value ψ, which could only be lower in the case of negative values. The highest score was also determined for parameter E5, as the detail is simple to construct, and the plastering and the reinforced concrete structure provide a good airtightness. The detail's scores from the aspect of durability and environmental impact are slightly lower. A lower score of E6 also results from the fact that the XPS was used whose parameters had high values for the LCA (i.e. GWP, AP, PEI). Deduction for durability (E7) is justified, since the detail is in contact with the ground which is constantly moist. Therefore, the installation of all protective layers is essential for the detail to function well throughout its lifetime. Weighting factors for the environmental and energy-efficiency evaluation are determined on the basis of the assessment which parameters are crucial to the foundation detail on XPS.

Since the seismic response may worsen due to the use of flexible XPS under the foundation slab, many points were deducted in the detail from the structural aspect (Table 5.3). With correct detailing, the solution with thermal insulation under the foundation slab can also be used in earthquake-prone areas, whereby sliding between thermal insulation boards, rocking on thermal insulation and extended fundamental period must be taken into account when modelling the global response. To determine the load-bearing capacity (K1) parameter, it was taken into account that the load-bearing capacity of the detail with XPS is worse in the event of an earthquake than if founded on the soil. The load-bearing capacity is greatly affected by the selection of material (XPS400 in our case). In preliminary studies, analyses showed that maximum compressive strength on the edge of the foundation slab may be exceeded (rocking may result in stress concentration) and irreversible compressive deformation may occur in XPS. It is difficult to determine the influence of the load-bearing capacity on the local level, which is why the influence of external factors (e.g. Z1—location and Z3—influence on the global analysis) should be taken into account. Similar is found for the influence of stiffness, which is reduced by the insertion of XPS under the foundation slab. A change in stiffness increases the structure's fundamental period, whereby higher buildings are subject to a greater change. To determine the score for K3, we were guided by material, as in a horizontal structural assembly, the selection of the material for the load-bearing structure is asymmetrical due to the

5.1 Building Connection Detail Between an Outer ...

foundation slab being placed on thermal insulation. A slightly lower score was given to parameter K4, since the load-bearing structure in the detail is not continuous all the way to the foundation base. The insertion of thermal insulation signifies a new material with poorer properties, violating the principle of uniformity. Parameter K5 scored well, as it does not contain any explicit eccentricity. Deduction for parameter K6 was established on the basis of the fact that irreversible compressive deformation and (or) uncontrolled sliding of XPS under the foundation slab may occur in a strong earthquake. Since this is a load-bearing layer under the foundations of a building, it should be protected. If it is damaged, it is virtually impossible to replace it. The last parameter (E6) scored well for the concrete structure, but received deduction because the detail is in constant contact with moisture. Weighting factors on the second level were determined on the basis of the assessment of the importance of parameters established with a parametric study of various structures founded on thermal insulation.

On the basis of analyses carried out, the load-bearing capacity criterion (K1) is assessed to have proportionately greater weight, since the detail is placed on the building foundation which is crucial to the response of the whole building. The highest weighting factor may be attributed to the stiffness parameter (K2). The analysis showed that a change in stiffness brought on by the insertion of thermal insulation may indirectly change the fundamental period of the whole building, affecting the magnitude of seismic forces acting on the building. In relation to the first two structural parameters, the symmetry parameter (K3) carries lower weight, as it does not have a significant influence in the analysed type of a detail. A higher weighting factor may be attributed to the continuity of the load-bearing structure (K4), since inserting thermal insulation under the foundation slab may violate the principle of the continuity of the load-bearing structure to the foundation base with a good load-bearing capacity. The parameter of the eccentricity of the load-bearing structure in relation to its primary load-bearing structural axis (K5), and the parameter of connections between the primary and secondary load-bearing structures (K7) are among the structural parameters deemed to have a low impact. This may be substantiated with the fact that this is an underground structural detail, which is only possible with a reinforced concrete load-bearing structure. In such a case, shifts in the load-bearing structure in relation to the main load-bearing structural axis can be solved more easily, and it is easier to fix secondary structural elements on account of the continuous primary load-bearing structure. The capacity design method (K6) also has a higher weighting factor than the structural parameters, since the detail is placed where the building is fixed, where potential damage in an earthquake would result in a more difficult and expensive restoration than in other structural details on the envelope of the building, which are more easily accessible.

Weighting factors for the environmental and energy-efficiency evaluation are determined in a similar way. Energy analyses showed that energy indicators on the local level are strongly affected by thermal transmittance (E1). Therefore, the weighting factor of this parameter is proportionately increased. It was also shown that the underground solution makes it irrelevant whether thermal insulation is continuous at all locations, which is why the influence of E2 is reduced. A high weighting factor is also attributed to the thermal comfort and condensation criterion (E3), as the detail is composed of a walk-on floor slab, whose surface temperature is the most important for the well-being of users (see Table 4.2). Analyses of influence on the use of energy showed that the influence on energy use parameter (E4) is important for such a type of a detail, resulting in an increase of its weight. A lower weighting factor is attributed to the airtightness parameter (E5) and the life cycle assessment (E6). A lower weighting factor for E5 can be substantiated by the fact that the detail is under the ground, where airtightness is not a significant component. On the basis of the review of materials for thermal insulation under the foundation slab, it can be stated for E6 that, in the current practice, there is no material that would provide a high load-bearing capacity, insulation and low environmental impact at the same time. Therefore, parameter E6 will have a lower weighting factor in the selection of structural solutions for such a detail. The highest weighting factor considering all environmental and energy-efficiency parameters is attributed to durability and stability (E7). Since the detail is under the ground, where potential aging and damage of materials would result in a considerably more expensive replacement or restoration of the detail, a higher weighting factor is sensible.

The total score (average) of the environmental and energy-efficiency parameters before weighting stands at 4.86 (81.0%) and of technical and structural parameters at 3.71 (61.9%). Such average score would be achieved by the detail with an even distribution of weighting factors and considering neutral external factors, which is used to compare the influence of external factors and weighting factors on the final score. This influence is disregarded for the other analysed details, as it is not important for detail comparison. In the next step, external factors (the level of influence on the significance of the criteria) are determined to correct the weighting factors from the second step of evaluation (Table 5.3). Weighting factors (the percentage of increase or decrease) and the division of categories of individual external criteria must be adapted to earthquake-prone areas and climatic conditions (E1) during evaluation, the importance of a building (E2) must be evaluated, the effect on the global analysis of a building (E3) must be known, the complexity of construction (E4) must be assessed, and the location and amount of penetrations (E5), and the input of funds for earthquake resistance and energy efficiency of the detail (E6) must be determined. An example of defining external factors by individual parameters is shown in Table 5.4 and follows the boundary conditions of the building connection detail. The final evaluation results with included unevenly distributed weighting factors and external parameters are given in Table 5.5.

According to Table 5.5, the total average score at the end of evaluation (after taking into account the weighting factors and the influences of external parameters) stand at 5.08 (84.7%) for the environmental and energy-efficiency evaluation and 3.43 (57.2%) for the technical and structural evaluation. Weighting and taking into account external factors are barely noticeable in the comparison of final score, as score for individual examples differ by a maximum of five percent.

Figures 5.3 and 5.4 are graphic presentations of the evaluation results. The radial diagram shows that the environmental and energy-efficiency evaluation scored better than the technical and structural evaluation. The form of the so-called floral diagram

5.1 Building Connection Detail Between an Outer ...

Table 5.4 External parameters influencing the environmental and energy-efficiency, and technical and structural evaluation

External parameters	Environmental and energy-efficiency parameters							Technical and structural parameters						
	E1	E2	E3	E4	E5	E6	E7	K1	K2	K3	K4	K5	K6	K7
Z1	2	1.50	1.75	1.50	1	0.50	1.25	1.50	1.25	1	1	1.50	1.25	0.25
Z2	0.75	1	1.50	1.25	0.50	0.50	2	1	1	1	0.75	1	1.50	1
Z3	2	1	0.50	0.50	1.50	0.75	0	1	2	0.75	1.25	0.50	1.75	0
Z4	0.50	2	1.75	1	2	1.25	2	1.25	1	1.50	1	1.75	1	1.25
Z5	1.50	1.25	1	1	2	0	0.50	0.50	1	1.50	1.25	1.75	2	1.25
Z6	1.50	0.75	1	2	1	1.50	1.25	1.50	1	1	0.50	1	2	1.50
Average	1.38	1.25	1.25	1.21	1.33	0.75	1.17	1.13	1.21	1.13	0.96	1.25	1.58	0.88

Table 5.5 Final evaluation results with included unevenly distributed weighting factors and external parameters

Environmental and energy-efficiency parameters	E1	E2	E3	E4	E5	E6	E7
Achieved score (O_{Ei}, O_{Ki})	5	6	5	6	6	2	4
Corrected weighting factors (Γ_{Ei}, Γ_{Ki})	2.06	0.94	2.50	1.81	0.67	0.19	2.04
Norm. corrected weighting factors ($\overline{\Gamma_{Ei}}, \overline{\Gamma_{Ki}}$)	0.20	0.09	0.24	0.18	0.07	0.02	0.20
Weighted score (O_{Ei-U}, O_{Ki-U})	1.01	0.55	1.22	1.07	0.39	0.04	0.80
Technical and structural parameters	K1	K2	K3	K4	K5	K6	K7
Achieved score (O_{Ei}, O_{Ki})	4	4	4	2	5	3	4
Corrected weighting factors (Γ_{Ei}, Γ_{Ki})	1.69	2.42	0.56	1.68	0.31	1.98	0.22
Norm. corrected weighting factors ($\overline{\Gamma_{Ei}}, \overline{\Gamma_{Ki}}$)	0.19	0.27	0.06	0.19	0.04	0.22	0.02
Weighted score (O_{Ei-U}, O_{Ki-U})	0.76	1.09	0.25	0.38	0.18	0.67	0.10

shows which parameters were not taken into account in the design of structural details. In this way, we can immediately see that virtually all parameters in the environmental and energy-efficiency evaluation were assessed as well, apart from parameter E6 which describes environmental impact. As mentioned, the weighting factor of this parameter was reduced. A very low weighting factor is also reflected in the value of the standard deviation, which is rather high (42.2%). A standard deviation may be used to recognise during evaluation whether any of the parameters significantly deviates from the average score. Therefore, this value can also be an indicator of balance of the consideration of all parameters of each evaluation. It can be stated for the specific detail that more attention should be paid to the development of materials and production processes with a low environmental impact. The technical and structural evaluation shows that the detail is assessed as satisfactory in most cases, but should be improved to produce a better response of the structure to seismic action. For this purpose, the following measures are recommended: selecting a thermal insulation material with a higher compressive strength, the use of

116 5 Case Study: Using Methodology to Assess the Selected Details

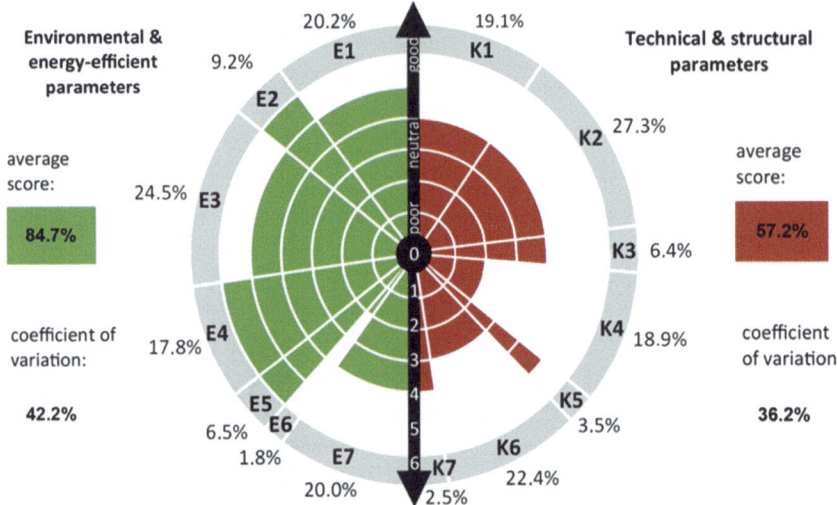

Fig. 5.3 Evaluation results of the building connection detail between the RC foundation slab and an RC outer wall presented in a radial diagram

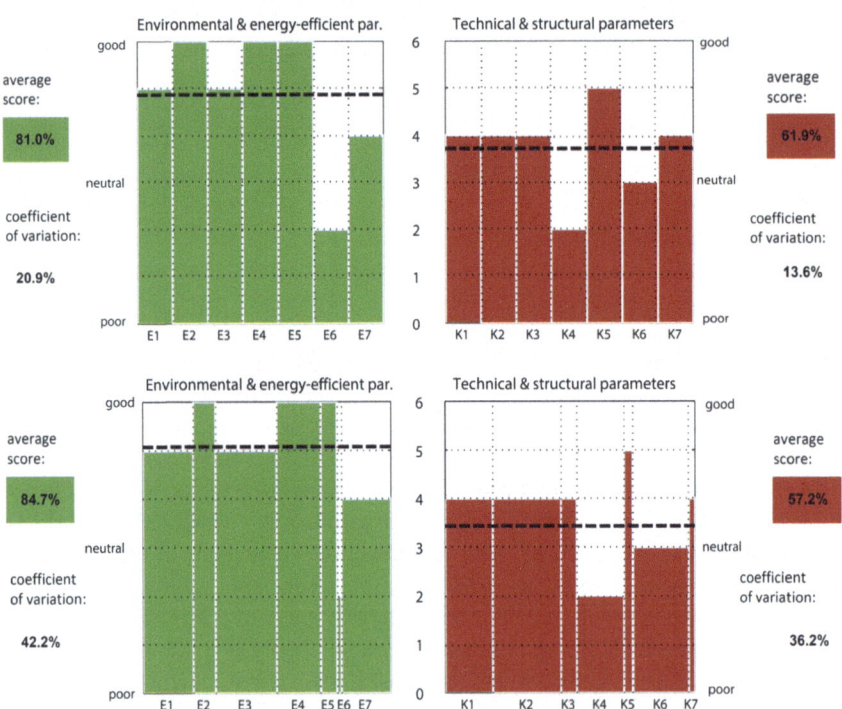

Fig. 5.4 Evaluation results of the building connection detail between the RC foundation slab and an outer wall presented for uniformly (top) and unevenly distributed weighting factors (bottom)

additional horizontal stoppers and vertical restrainers to limit sliding and rocking of the building, installation protection, etc. (e.g. see the description in (Azinović et al. 2015b, 2016). Given the criteria for final evaluation in Sect. 4.6, the detail is assessed as "limited use".

Figure 5.4 is shown particularly to facilitate the comparison of the influence of weighting and external factors, and contains diagrams with evenly and unevenly distributed weighting factors. The horizontal dashed line showing the average results can also be used to comment on the results from the column charts. The greater the shift in both lines that show the average, the more unbalanced the principles of the environmental and energy-efficiency, and technical and structural evaluation. To design each detail, both parts of evaluation should be as balanced as possible, which may be graphically shown by aligning the horizontal lines that show the average. In the concrete case, the difference between both parts of evaluation increases if weighting and external factors are taken into account. In addition to changing average final scores, on which the impact is not significant in the concrete case (max. 5%), the application of weighting and external factors also affects standard deviation. The latter is reflected in the significantly changed shape of the graph, which is an important piece of additional information for the assessor. More results for various foundation details are shown in the appendix, which consider the evaluation criteria equally. Even distribution of weighting and external factors is used for simplification.

5.2 Building Connection Detail Between the Load-Bearing Balcony Structure and an Outer Wall

This section discusses the building connection between the reinforced concrete (RC) balcony slab and an outer masonry wall (Fig. 5.5). The thermal bridge that occurs due to the penetration of the RC slab is reduced or eliminated by inserting a precast load-bearing thermal insulation element (1*). The heat transfer simulation results and the numerical analysis of the seismic response are used to evaluate the detail according to the proposed methodology (Azinović et al. 2014, 2015a). The thickness of the load-bearing thermal insulation element (LBTIE) is 8 cm and its thermal conductivity stands at $\lambda = 0.12$ W/(m K). It is assumed that the precast element contains the longitudinal reinforcement only in the upper part of the slab. Such elements were examined in detail in previous studies (Ge et al. 2013; Goulouti et al. 2014), making it easier to assess individual evaluation parameters based on those results. The detail is composed of an outer masonry wall insulated with mineral wool ($U = 0.11$ W/(m^2 K)), an interstorey slab and a RC balcony slab. All data on the analysed structural assemblies are provided in Table 5.6.

Figure 5.6 shows the results of the numerical simulation of heat transfer through the analysed connection detail between the RC balcony slab and LBTIEs. The left part of Fig. 5.6 reveals how precast elements successfully prevent a thermal bridge. Surface temperatures are very high and are not lower than $\theta_{si, \text{min.}} = 18.5$ °C in any

Fig. 5.5 Analysed building connection detail of the RC balcony slab with LBTIE

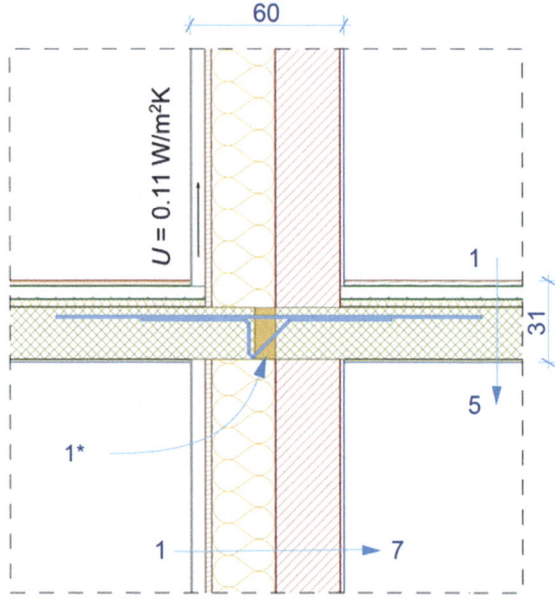

Table 5.6 Composition of structural assemblies for the building connection detail between the RC balcony slab and LBTIE

Outer masonry wall Thermal transmittance U = 0.11 W/m²K			Reinforced concrete interstorey slab		
No.	Material	T (cm)	No.	Material	T (cm)
1	Fibre cement boards	0.8	1	Flooring—wooden parquet	1.5
2	Ventilated layer	5	2	Concrete screed	5
3	Wind barrier (felt, geotextile)	0.1	3	Acoustic mineral wool	3
4	Plywood	2.4	4	Reinforced concrete load-bearing slab	20
5	Stone wool in timber substructure	25	5	Finishes	–
6	Hollow brick (Poroblock 29/25)	25			
7	Cement mortar	1.5			

part of the detail. The analysis of the temperature range gives rise to the finding that the thickness of the thermal envelope is reduced at the location of the LBTIEs, which does not significantly affect the inner surface temperature. On the other hand, a certain effect on the use of energy must be expected, since heat flow is still more intensive at this critical part of the envelope despite the use of thermal insulation

5.2 Building Connection Detail Between the Load-Bearing ...

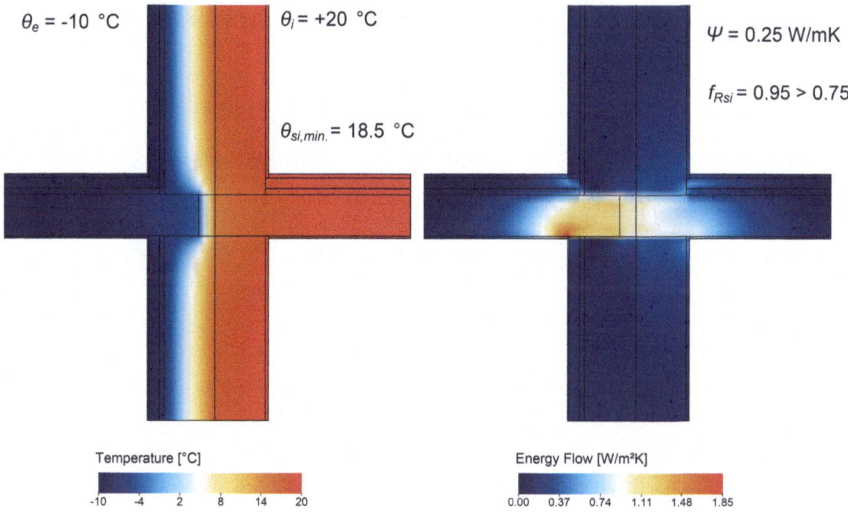

Fig. 5.6 Results of the numerical simulation of heat transfer through the building connection detail between the RC balcony slab and LBTIEs

elements (the right side of Fig. 5.6). The linear thermal transmittance coefficient of the analysed detail is $\psi = 0.25$ W/(m K). This could be improved with a thicker load-bearing thermal insulation element or better precast elements with a lower thermal conductivity.

As shown by the heat transfer simulation results, a thermal bridge resulting from the penetration of the RC slab may be reduced by inserting a LBTIE. This is taken into account in the detail evaluation with a high total score from the environmental and energy-efficiency aspect (Table 5.7). The outer wall structural assembly has a good thermal insulation with a low thermal transmittance that corresponds to the PH standard. Therefore, the detail receives the highest score for parameter E1. Deduction from the full score was assumed for parameter E2, as the thickness of the LBTIE is lesser than thickness of thermal insulation on the outer masonry wall. In addition, the thermal conductivity of this element is higher than the conductivity of the remaining thermal insulation. The analysis of the temperature range showed high surface temperatures, which ensure thermal comfort, and prevent condensation and mould. To assess the effect on the use of energy, it must be taken into account that the penetration of a cantilever slab is among the most complex thermal bridges, since a combination of a structural and geometrical thermal bridge usually appears at this location. We believe that the effect on the use of energy could reduce if high-quality load-bearing thermal insulation elements were used. Therefore, the score for parameter E4 is slightly lower than the full score. In the case of wider balconies, these thermal bridges are still recommended to be considered in the calculation of energy use in the building. From the aspect of airtightness (E5), the detail was well assessed because of the simplicity of construction, and because the plastering and RC

Table 5.7 Final evaluation results of the building connection detail of the RC balcony slab with LBTIEs

		Score	Selected weighting factors (weights)	Corrected weighting factors (influence of external factors)	Share of weighted score
Environmental & energy-efficiency parameters	E1	6	0.75	0.75	0.47
	E2	5	1.75	1.75	0.92
	E3	5	2	2	1.05
	E4	5	2	2	1.05
	E5	4	1.25	1.25	0.53
	E6	2	1	1	0.21
	E7	5	0.75	0.75	0.39
Technical & structural parameters	K1	3	0.75	0.75	0.27
	K2	2	1.50	1.50	0.36
	K3	2	2	2	0.48
	K4	2	2	2	0.48
	K5	4	0.50	0.50	0.24
	K6	3	0.25	0.25	0.09
	K7	4	1.25	1.25	0.61

structure provide good airtightness. Certain deduction was made on account of openings (windows, the balcony door), which are indispensable in the balcony structure detail. The detail was strictly assessed from the aspect of its environmental impact (E6) resulting from the used materials (RC, XPS and brick). As the detail is tested and frequently used in practice, there was no significant deduction for E7.

On the other hand, it is more poorly assessed from the technical and structural aspect and must be improved to be used in earthquake-prone areas. To determine the score for parameter K1, it was taken into account that the load-bearing capacity of the detail is good under vertical static loads (experimentally tested), while the reinforcement in the lower part of the cross-section, which would provide the load-bearing capacity during the cantilever uplift in the event of stronger (vertical) seismic action, was not taken into account. In addition, reduced stiffness (parameter E2) of the RC balcony cantilever slab where fixed must be considered, as it increases deflections also in the case of vertical static loads. Therefore, designers must pay special attention to the limitation of the maximum length of the fixed cantilever to meet the controls of the maximum deflection in the serviceability limit state. During the assessment of parameter K3, it was found that the detail is not symmetrical (the longitudinal steel reinforcement is not placed on the compressive and tensile side). Since the load-bearing structure is not completely continuous (load-bearing thermal insulation elements interrupt the RC cantilever slab), a significant deduction was made also for parameter K4. To determine the score for K5, it was taken into

5.2 Building Connection Detail Between the Load-Bearing ...

account that there is no explicit eccentricity, except in the precast thermal insulation element. The principle of the capacity design method is not taken into account in the analysed detail. However, K6 of the detail is relatively well assessed, as it is not critical to the global stability of the building. The contact between secondary (non-)load-bearing elements and the primary load-bearing structure is generally not problematic, but this detail has an increased thickness of thermal insulation and a ventilated layer. In addition, load-bearing thermal insulation elements are placed at the location of maximum bending moment (fixed end of the cantilever), where a load-bearing substructure for the balcony door must be ensured. For these reasons, the score for parameter K1 is suitably lower.

Weighting factors on the second level of the evaluation were determined on the basis of the assessment of the importance of parameters established with a structural and energy analysis of precast cantilever elements (see Table 5.7). The influence of external factors is neglected (default value 1) for simplification. The analyses showed that the load-bearing capacity (K1) was not relevant in most cases on account of high safety factors for vertical static loads, which is why parameter K1 is attributed a lower weighting factor. On the other hand, an increase in stiffness (K2) and the asymmetric placement (K3) of steel reinforcement significantly affect the response of the cantilever structure due to the insertion of more flexible load-bearing thermal insulation elements. Therefore, parameters K2 and K3 are attributed a higher weighting factor. The analyses also showed that the discontinued connection of the load-bearing structure has a great influence on the structural response. For this reason, parameter K4 is attributed a higher weighting factor. The parameters of the eccentricity of the load-bearing structure in relation to the axis (K5) and the capacity design method (K6) proved to have a very low impact on the response. This is a detail, which generally does not significantly affect the global seismic safety of the load-bearing structure, but is basically accessible on the building envelope. Therefore, K5 and K6 do not have a significant effect as in the foundation detail. A higher weighting factor is attributed to parameter K7, as this is a detail of the »+« type connection, in which the fixing of secondary (non-)load-bearing elements is more complex.

For the weighting factors of the environmental and energy-efficiency evaluation of the cantilever structure detail, the results of energy analyses were taken into account, indicating that the thermal transmittance of an outer wall has less influence on the energy properties of the detail than continuous thermal insulation (see the results in Fig. 3.16). Therefore, the weighting factor for parameter E1 is suitably lower than for parameter E2. Experience from practice, which prompted the emergence of precast thermal insulation elements, shows that the main problem of such a type of detail is its effect on the use of energy, and the occurrence of condensation and mould. To this end, the highest weighting factor is attributed to parameters E3 and E4. Greater influence is also assumed for airtightness (E5), as this is a »+« type detail, in which ensuring an airtight plane is difficult. No specifics are assumed for the life cycle assessment (E6). That is why an even distribution of weighing factors is maintained. A lower weighing factor is assumed for durability and stability (E7), as this is a frequently used detail on the envelope exposed to usual external actions. In addition, the detail is accessible, and can be replaced in the event of damage (e.g. caused by ultimate limit state) more simply and at lower costs than, for example, the foundation detail.

The diagrams in Fig. 5.7 present the final evaluation of the fixed RC cantilever with LBTIEs. Due to the potential cantilever uplift in the event of a strong earthquake and the lack of lower steel reinforcement the final detail score is unsuitable.

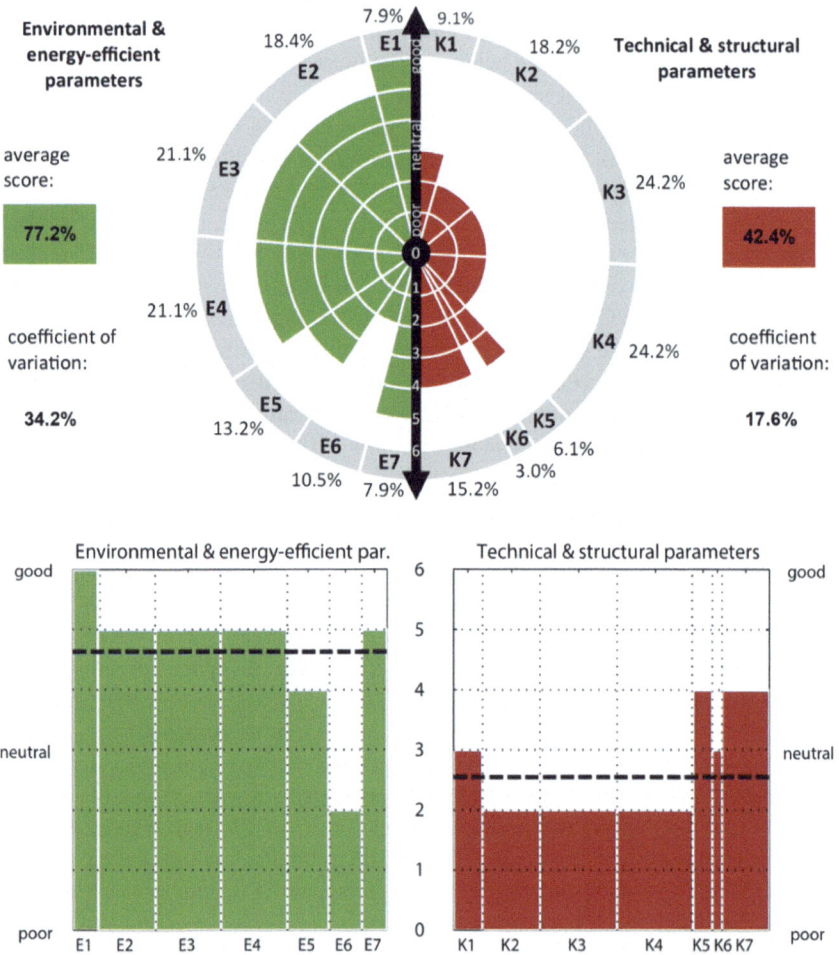

Fig. 5.7 Evaluation results presented in radial and column diagrams for the building connection detail between the RC balcony slab and LBTIEs

5.3 Building Connection Detail Between an Outer Wall and the Unheated Basement

This section addresses the building connection detail between an outer masonry wall and the unheated basement (Fig. 5.8). From the aspect of preventing thermal bridges, the detail is very complex, as the intensive heat flow towards the unheated basement and outdoor air must be prevented. To reduce a thermal bridge, designers used to decide on the solution to extend thermal insulation to the basement wall. This improves the temperature profile, but does not eliminate the thermal bridge towards the unheated basement. There is basically no solution to completely eliminate a thermal bridge without interrupting the load-bearing structure. With skeleton load-bearing structures, a thermal bridge in columns should be accepted, since interruptions in this section of the primary load-bearing structure are not admissible either from the aspect of the load-bearing capacity under vertical static loads or the aspect of seismic action. However, as presented in Sect. 3.2, an insulation base block can be used to interrupt or reduce a thermal bridge in certain masonry load-bearing structures. Such solutions are possible mainly due to lower compressive stresses in primary load-bearing structures in comparison with skeleton load-bearing structures. Solutions with base insulation blocks at the location where they are fixed stem particularly from earthquake-non-prone areas, where the only structural control is to ensure sufficient compressive strength for these blocks. However, additional loads, which may occur at this location of the load-bearing structure (where outer walls are fixed) must be considered in earthquake-prone areas.

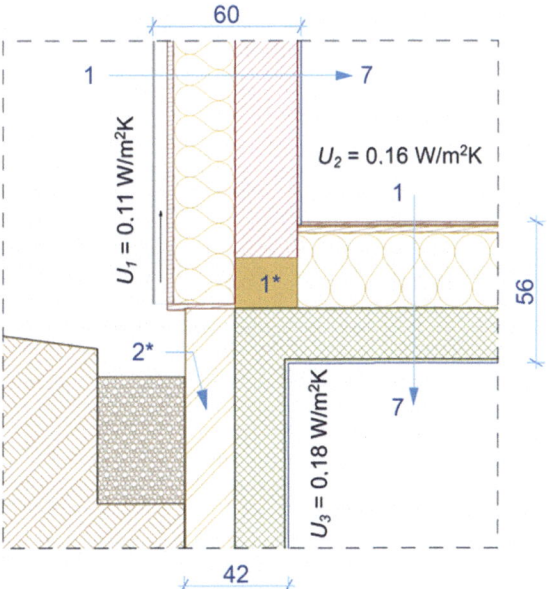

Fig. 5.8 Analysed building connection detail between an outer masonry wall and the RC unheated basement

In the selected evaluation case, the following structural assemblies were analysed (Table 5.8): (i) an outer masonry wall insulated with mineral wool ($U_1 = 0.11$ W/(m² K)); (ii) the RC interstorey slab insulated with perlite aggregate ($U_2 = 0.16$ W/(m² K)); and (iii) a basement wall from concrete hollow bricks insulated with XPS ($U_3 = 0.18$ W/(m² K)). A thermal bridge that occurs due to the contact between an outer wall and the unheated basement is interrupted by an insulation base block (1*) with vertical (longitudinal) thermal conductivity $\lambda = 0.13$ W/(m K). In the concrete case, it is assumed that the insulation base block is made of autoclaved aerated concrete with greater porosity to achieve lower thermal conductivity of the base block. On the other hand, such autoclaved aerated concrete masonry is characterised by lower compressive strength. More options of base insulation blocks are provided in Sect. 3.2 (see, for example, Table 3.4). The thermal insulation on the basement wall is installed at least one metre below the underground section (2*).

Figure 5.9 shows the results of the numerical simulation of heat transfer for a structural detail with an insulation base block. The analysis of the detail took into account the simplification that the temperature of the unheated basement is equal to the outdoor temperature (θ_e). The performed numerical analysis enables us to distinguish two values of linear thermal transmittance: (i) for heat flow towards outdoor air and (ii) for heat flow towards the unheated basement. Both values are very low and depend particularly on the vertical thermal conductivity of the insulation base block (see, for example, Figs. 3.11 and 3.12). The results also showed that surface temperatures are very low and do not fall to values lower than $\theta_{si,\,min.} = 17.1$ °C. This means that condensation and mould cannot occur in any part of the detail, and the desired thermal comfort is achieved, as temperatures are close to the

Table 5.8 Composition of structural assemblies for the connection detail between an outer masonry wall and the RC unheated basement

Outer masonry wall Thermal transmittance $U_1 = 0.11$ W/(m² K)			Interstorey slab between basement and ground floor Thermal transmittance $U_2 = 0.13$ W/(m² K)		
No.	Material	T (cm)	No.	Material	T (cm)
1	Fibre cement boards	0.8	1	Flooring—wooden parquet	1
2	Ventilated layer	5	2	Soft foam, PE	0.5
3	Wind barrier (felt, geotextile)	0.1	3	Vapour barrier, Sd = 2 m*	0.1
4	Plywood	2.4	4	OSB board	2.4
5	Stone wool in timber substructure	25	5	Perlite aggregate	30
6	Hollow brick (Poroblock 29/25)	25	6	Stone wool	1
7	Cement mortar	1.5	7	Reinforced concrete slab	20

*Sd factor, expressed in metres, represents the resistance of a vapour barrier in comparison with the resistance produced by the equivalent thickness of an air layer

5.3 Building Connection Detail Between an Outer Wall …

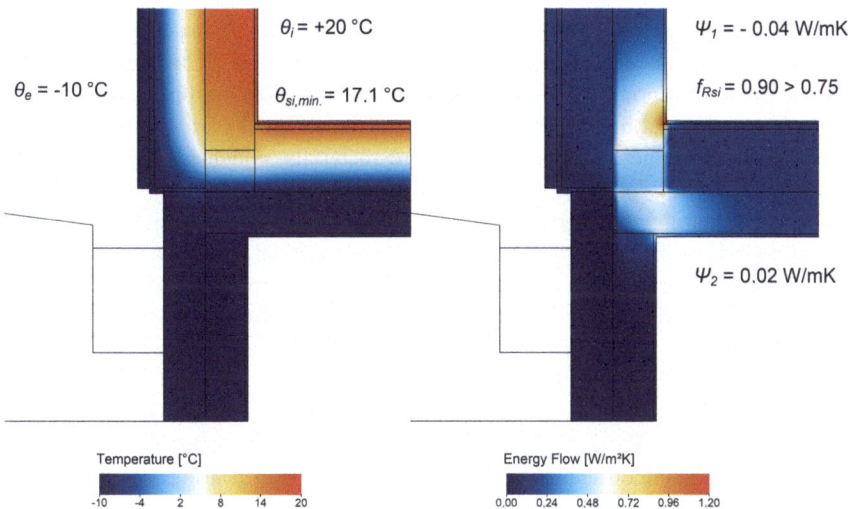

Fig. 5.9 Results of the numerical simulation of heat transfer through the building connection detail between a masonry wall and the unheated basement separated by base insulation blocks Ψ_1…linear thermal transmittance in direction of the external air; Ψ_2…linear thermal transmittance in direction of the unheated basement

indoor air temperature (θ_i). The right side of the figure indicates that heat flow is greater towards the unheated basement, which is supported by the values of linear thermal transmittance ($\psi_2 > \psi_1$).

Due to low thermal transmittance, favourable temperature range, the environmental and energy-efficiency parameters of the detail received a good total score. To assess parameter E1, it was taken into account that the thermal transmittance of both structural assemblies was lower than the PH standard requirements, but higher than recommended by current regulations in most Central European regions (i.e. 0.10 W/(m² K)). From the aspect of the continuity of thermal insulation (E2), the score is reduced due to a change in the thickness of thermal insulation at the location of the insulation base block. In addition, the vertical thermal transmittance of the base block is poorer than of the selected thermal insulation. As shown by the analysis results of the temperature range and heat flow, the detail has little influence on the use of energy and thermal comfort, making the scores for E3 and E4 suitably high. As the detail is relatively complex (many different materials, connections, etc.), certain airtightness requirements (simplicity, the continuity of the plane, airtightness cannot be ensured, etc.) are violated, reducing the score for E5. The scores for E6 and E7 are also slightly lower due to the used non-renewable materials (e.g. XPS, concrete, etc.). Therefore, a greater environmental impact may be expected. In the durability assessment, the installation complexity of a ventilated wooden façade was taken into account. To ensure the durability of a wooden façade and its thermal insulation role, numerous protective layers (e.g. wind barrier, UV protection, etc.) must be installed.

A thermal bridge is solved by an insulation base block, whose properties are generally poorer than of reinforced concrete and masonry walls. This means a weakening of the load-bearing structure at the location where the masonry wall is fixed. Due to better thermal insulation, most base insulation blocks are characterised by lower density than the load-bearing structure materials, which is why their load-bearing capacity for compression and shear is usually worse. Such a detail affects the total load-bearing capacity of masonry structures and must be considered in structural earthquake resistance evaluation. Since this is a critical detail in a structure, it is poorly assessed from the structural aspect (Table 5.9). Due to its porous structure, the used autoclaved aerated concrete is characterised by lower compressive and tensile strength. Therefore, the detail is assessed as unsuitable for earthquakes (see assumptions in Sect. 3.2). Since the insulation base block is also characterised at this location by lower stiffness than of the selected load-bearing structure materials, it is attributed a very low score for K1 and K2. To determine the score for parameter K3, the fact that the detail is not symmetrical in view of its horizontal load-bearing axis was taken into account, since it constitutes a change in the load-bearing structure material. The continuity of the vertical load-bearing structure is violated for the same reason, reducing the score for K4 (the masonry wall is not continuously connected with the RC base). It was taken into account for K5 that the masonry wall with protective layers is slightly displaced above the RC basement wall. When assessing consideration of the capacity design method (K6), it must be noted that this is a critical detail

Table 5.9 Final evaluation results for the building connection detail between a masonry wall and the unheated basement

		Score	Selected weighting factors (weights)	Corrected weighting factors (influence of external factors)	Share of weighted score
Environmental & energy-efficiency parameters	E1	5	1	1	0.71
	E2	4	1	1	0.57
	E3	6	1	1	0.86
	E4	5	1	1	0.71
	E5	4	1	1	0.57
	E6	3	1	1	0.43
	E7	4	1	1	0.57
Technical & structural parameters	K1	2	1	1	0.29
	K2	1	1	1	0.14
	K3	2	1	1	0.29
	K4	0	1	1	0
	K5	4	1	1	0.57
	K6	0	1	1	0
	K7	3	1	1	0.43

5.3 Building Connection Detail Between an Outer Wall ...

at the location where the load-bearing structure is fixed to the stiff RC basement. It was assessed that damage is very likely to occur at the unfavourable location where the wall is fixed due to the weakened load-bearing structure at this location. For this reason, the detail received the lowest possible score (0) for parameter K6. The score for K7 was also reduced because the increased thickness of thermal insulation and the ventilated layer make the fixing of secondary (non-)load-bearing elements difficult.

Unlike in Sects. 5.1 and 5.2, in which concrete cases were used to show the evaluation method and calculation procedure in the proposed methodology, external and weighting factors are not taken into account in this case, resulting in a more concise display of the results. Only final values (Table 5.9), and radial and column diagrams (Fig. 5.10) are displayed. The graphic display of the results shows a distinctly asymmetrical ratio between the scores for the detail's structural parameters, which is evident from the difference between the average scores. From the structural aspect, the diagram is insufficient, while it is suitable for energy-efficient buildings from the environmental and energy-efficiency aspect. The detail was strictly assessed from the technical and structural aspect, as no experimental data are available for masonry with base insulation blocks. As mentioned in the chapter on assumed influence of base insulation blocks on seismic safety, the problem of base insulation blocks from the structural aspect has not yet been addressed in professional and scientific literature. In this case, the conservative values for the parameter assessment must be adopted in the proposed methodology, since scores can only be made on the basis of approximations and assumptions (e.g. the ratio between the load-bearing capacity of base insulation blocks and the remainder of the load-bearing structure, general engineering principles, etc.). Since base insulation blocks are mainly used in non-earthquake-prone areas, better thermal insulation is expected along the cross-section of the masonry wall, as reinforced concrete vertical ties are not required. In the case of masonry load-bearing structures in earthquake-prone areas, masonry must be additionally connected with horizontal and vertical reinforced concrete ties. It must be noted that a thermal bridge could still occur at the location, which would lead to a lower total environmental and energy-efficiency score. The latter may be taken into account in the proposed evaluation methodology with external parameters (e.g. Z1 and Z3).

Based on the scores according to the proposed evaluation methodology and assumptions in Sect. 3.2, it may be concluded that the detail's load-bearing capacity and ductility will be poorer, which is why it is not recommended in earthquake-prone areas. For better predictions of the response to seismic action and the use of such solutions, masonry with base insulation blocks should be experimentally tested and specific instructions for modelling, designing and executing in earthquake-prone areas must be provided. More examples of structural details with base insulation blocks are provided in the appendix.

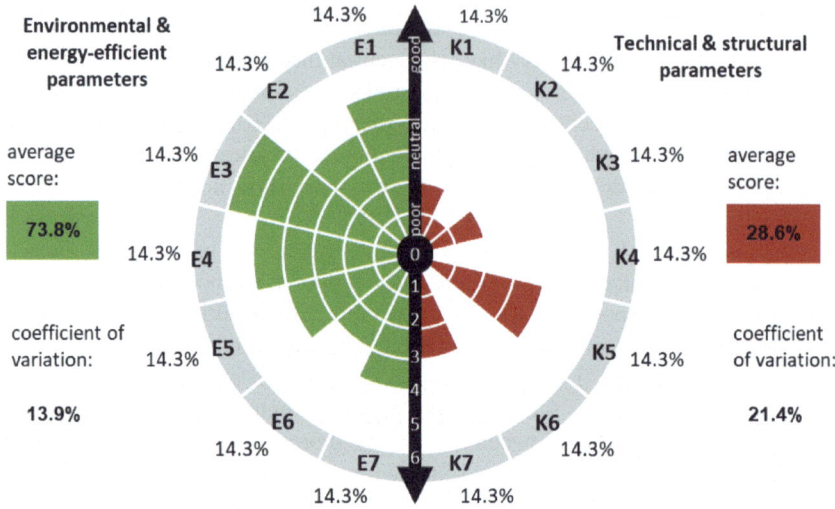

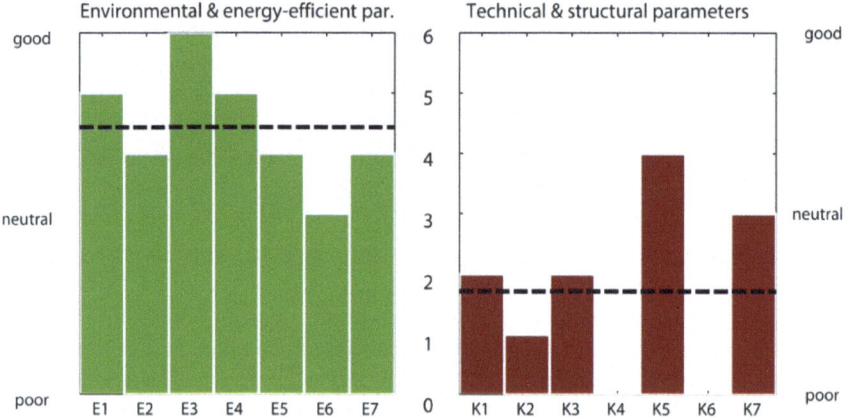

Fig. 5.10 Evaluation results presented in radial and column diagrams for the building connection detail between an outer masonry wall and the unheated basement separated by base insulation blocks

5.4 Building Connection Detail Between an Outer Wall and the Roof

This section addresses the building connection detail between the RC flat roof and an RC outer wall with thermal insulation on the internal side (Fig. 5.11). The roof slab is extended with a cantilever over the outer wall to provide a RC canopy overhang. In the previous sections, details suitable for energy-efficient buildings were analysed. This section, however, addresses a detail that is not confirmed according to the PH standard. Therefore, greater influence on the use of energy than of previous details

5.4 Building Connection Detail Between an Outer Wall and the Roof

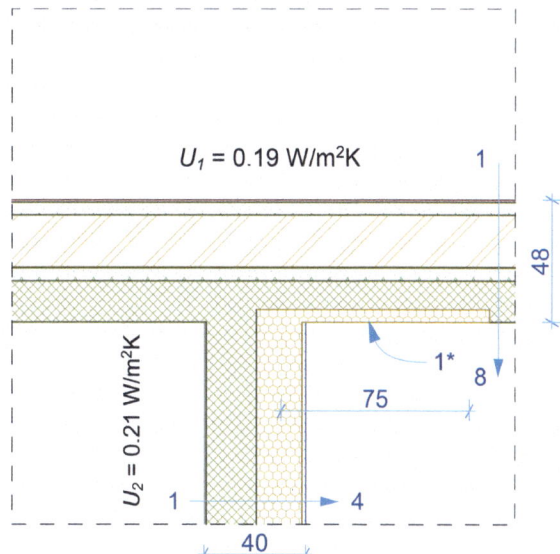

Fig. 5.11 Analysed building connection detail between an RC outer wall and the RC flat roof with an overhang

may be expected, as the used structural assemblies have higher thermal transmittance than required by the PH standard. Also disadvantageous for the detail from the energy aspect is that a thermal bridge occurs at the location of the RC structure connection, since thermal insulation at this location cannot be continuous. The thermal bridge in the analysed detail can be partially eliminated by installing thermal insulation at the location where the RC roof slab is fixed (1*). The length of thermal insulation from the location where the RC roof slab is fixed is 75 cm and its thickness is 5 cm. The installation of thermal insulation means that the dimension of the RC load-bearing structure is reduced in this part to maintain a flat surface for the final treatment of the ceiling. The detail is composed of an RC flat roof slab with pitched concrete and thermally insulated with XPS ($U_1 = 0.19$ W/(m² K)), and an RC outer wall thermally insulated with EPS ($U_2 = 0.21$ W/(m² K)). More information on the composition of structural assemblies is in Table 5.10.

Figure 5.12 shows the heat transfer simulation results for the analysed detail of a flat roof with an RC canopy overhang. On the left side of the figure, we can see that the temperature range of the structural assembly is not very favourable. In practice, walls with thermal insulation on the internal side are not desired, as the temperature drop at this location of thermal insulation is substantial. This means that condensation or the dew point could occur in thermal insulation. In theory, a vapour barrier would have to be installed on the internal side before thermal insulation. This would result in more complex details, since a vapour barrier and other protective layers require sealing to support the basic role of thermal insulation. Despite thermal insulation on the internal side, the proposed measure (decreasing the RC slab thickness) resulted in much higher surface temperatures in the detail. The minimum surface temperature in the detail stands at $\theta_{si,\,min.} = 16.1\,°C$.

Table 5.10 Composition of structural assemblies for the building connection detail between an RC outer wall and the RC flat roof with an overhang

Walk-on RC roof, thermal transmittance $U_1 = 0.19$ W/(m² K)			Reinforced concrete outer wall, thermal transmittance $U_2 = 0.21$ W/(m² K)		
No.	Material	T (cm)	No.	Material	T (cm)
1	Ceramic tiles	1	1	Silicate thin-layer plastering	–
2	Concrete screed	5	2	Reinforced concrete outer wall	20
3	Secondary waterproofing	–	3	EPS (expanded polystyrene)	20
4	XPS (extruded polystyrene)	20	4	Finishes	–
5	Bitumen cardboard (primary waterproofing)	1			
6	Pitched concrete	5			
7	Reinforced concrete roof slab	16			
8	Finishes	–			

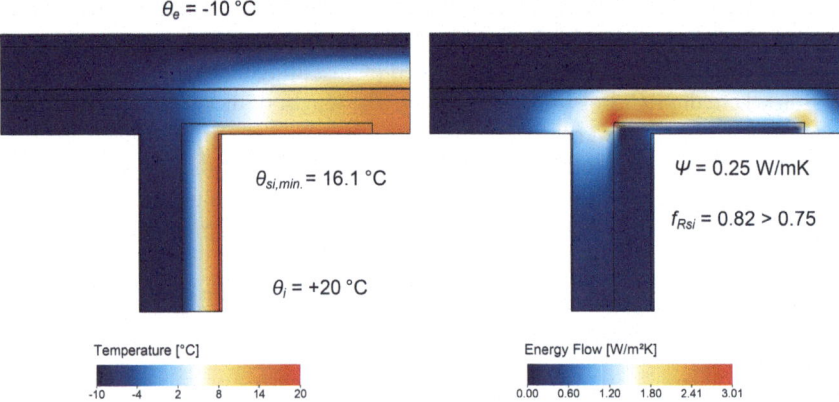

Fig. 5.12 Results of the heat transfer numerical simulation for heat transfer through the building connection detail between the RC outer wall and the RC flat roof with an overhang

On the other hand, the measure of adding thermal insulation on the account of reducing the size of the load-bearing structure only reduces the thermal bridge. As seen from the right side of Fig. 5.12, intensive heat flow is interrupted in the corner, where thermal insulation is installed. However, heat transfer can still be expected in other sections of the load-bearing structure (around thermal insulation). Since thermal insulation is not continuous, the linear thermal transmittance coefficient still stands

5.4 Building Connection Detail Between an Outer Wall and the Roof

at $\psi = 0.25$ W/(m K). From the aspect of preventing thermal bridges, the only option in the analysed case is to interrupt the load-bearing structure. Theoretically, there are two options: (i) installing thermal insulation around the overhang and interrupting the load-bearing wall or (ii) interrupting the load-bearing roof structure. None of these solutions is desired in earthquake-prone areas from the structural aspect.

As shown in the analysis of the temperature range, the thermal bridge is not completely eliminated in the analysed detail, making the latter only conditionally useful from the aspect of the environmental and energy-efficiency parameters (Table 5.11). The detail is poorly assessed for almost all parameters. To assess E1, the fact that the thermal transmittance of the outer wall and the roof is higher than required by the PH standards and very close to the minimum value from the common technical guidelines used in Central Europe, which is not appropriate for modern energy-efficient buildings, was taken into account. To assess E1, the fact that thermal insulation is not continuous along the envelope was considered. The continuity is partly saved by the reduced thickness of the load-bearing structure in the critical region, resulting in slight bonus for the final score. In view of the input project data, condensation and mould cannot occur (parameter E3), but the solution is poorer from the aspect of condensation, as thermal insulation is located on the internal side (a vapour barrier must be installed). The score for parameter E3 received a bonus for the detail's location on the roof (the effect on thermal comfort is lower). The influence on the use of energy is significant due to high linear thermal transmittance

Table 5.11 Final evaluation results for the building connection detail between an outer wall and the roof

		Score	Selected weighting factors (weights)	Corrected weighting factors (influence of external factors)	Share of weighted score
Environmental & energy-efficiency parameters	E1	3	1	1	0.43
	E2	2	1	1	0.29
	E3	2	1	1	0.57
	E4	2	1	1	0.29
	E5	5	1	1	0.71
	E6	2	1	1	0.29
	E7	5	1	1	0.71
Technical & structural parameters	K1	1	1	1	0.14
	K2	2	1	1	0.29
	K3	3	1	1	0.43
	K4	3	1	1	0.43
	K5	4	1	1	0.57
	K6	3	1	1	0.43
	K7	4	1	1	0.57

($\psi = 0.25$ W/(m K)). It was additionally taken into account that the detail runs along the edge of the roof. Therefore, the score for E4 is suitably low. Parameter E5 was well assessed because of the simplicity of construction, and because the plastering and brick provide good airtightness. The environmental and energy-efficiency score for E6 of this detail is low due to the used materials, i.e. XPS and RC. Regarding durability (parameter E7), it was taken into account that the detail was tested and simple to build.

In structural terms, the detail was poorly assessed (Table 5.11), as the thermal bridge was reduced by decreasing the dimension of the load-bearing RC roof structure. This made the detailing of the contact in earthquake-prone areas difficult, which requires, in critical regions, higher ratio of steel reinforcement, a symmetrical set-up of steel reinforcement, suitable stiffness and the load-bearing capacity of the contact. In addition, weakening must be considered in the capacity design method principle, the global model of the building, and the seismic analysis. To determine the parameters for the technical and structural aspect, experiential engineering approach was applied, which follows the principles of Eurocode 8. To determine the score for the load-bearing capacity (K1), the assumptions of a lower load-bearing capacity under vertical static and cyclic loads (due to weakening in the roof), reduced load-bearing capacity (potential damage to the RC roof in the case of substantial loads), and the critical region made difficult due to the reduced dimensions of the load-bearing structure were taken into account. A similarly low score was attributed to K2, since stiffness is expected to reduce due to an interruption in thermal insulation. Such a weakening must be taken into account in seismic analyses—a partially stiff joint is analysed. Parameter K3 was assessed as satisfactory. There was a deduction from the full score, as the detail is not symmetrical in view of the horizontal load-bearing axis resulting from the insertion of thermal insulation (reduced dimension of the load-bearing roof structure). From the aspect of continuity (parameter K4), it may be foreseen, on the basis of the known data, that the detail is continued vertically but not horizontally, as it is complicated to anchor reinforcement for the overhang. It is also difficult to ensure continued longitudinal steel reinforcement to the outer wall. A slightly lower score may be expected for K5 due to a shift in relation to the horizontal axis brought on by the insertion of thermal insulation. In the assessment of K6, it was taken into account that the detail is not critical in the global context of the building. There is also weakening in the load-bearing roof structure, and the vertical load-bearing structure is protected, which is favourable for the capacity design method parameter. Nevertheless, designers should foresee weakening in the load-bearing structure in the global seismic analysis. As already mentioned, the reduced thickness of the load-bearing structure in the critical region results in difficult fixing of secondary (non-) load bearing elements and the anchoring of the overhang. Therefore, the score for K7 is slightly lower (Fig. 5.13).

Overall, we believe that the evaluated detail should be redesigned or replaced. Several options for the building connection detail between an outer wall and the roof are shown in the appendix.

5.4 Building Connection Detail Between an Outer Wall and the Roof

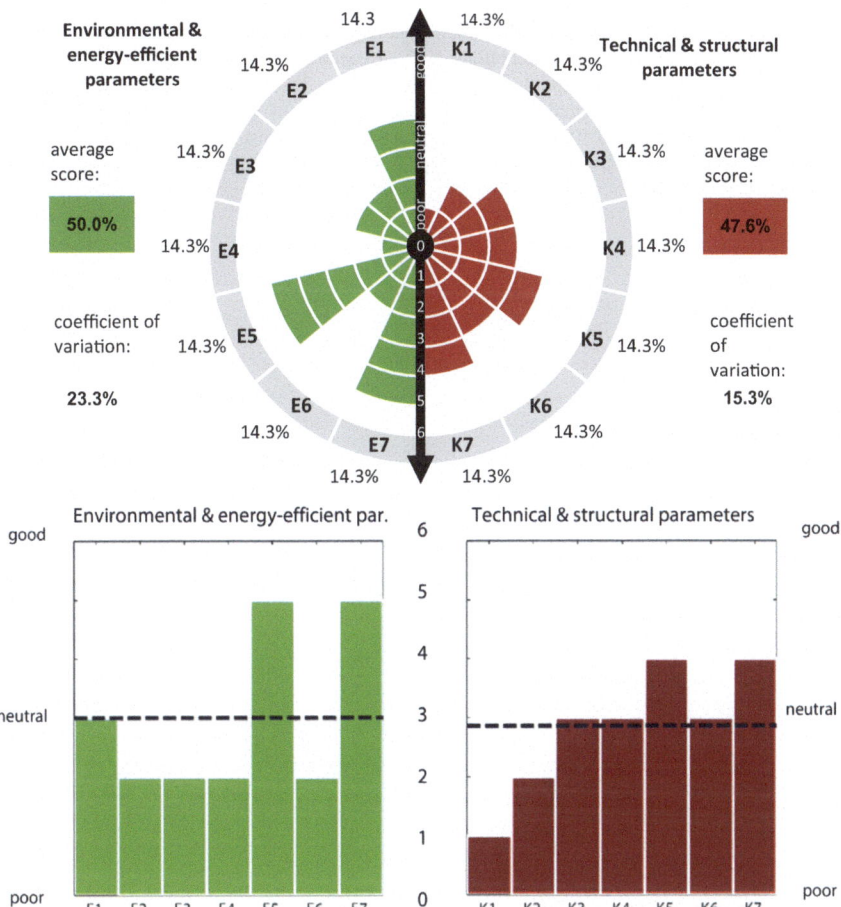

Fig. 5.13 Evaluation results presented in radial and column diagrams for the building connection detail between an RC outer wall and the RC flat roof with an overhang

References

Azinović B, Koren D, Kilar V (2014) Seismic safety of the precast balcony cantilever elements for prevention of thermal bridges. Architec Res 2014:25–33

Azinović B, Kilar V, Koren D (2015a) Erdbebensicherheit vorgefertigter wärmegedämmter Stahlbeton-Konsolenelemente. Bauingenieur 90:489–499

Azinović B, Koren D, Kilar V (2015b) Sliding isolation system for controlled response of passive houses founded on thermal insulation. In: 14th World conference on seismic isolation, energy dissipation and active vibration control of structures. Anti-Seismic Systems International Society (ASSISi), San Diego, USA

Azinović B, Kilar V, Koren D (2016) Energy-efficient solution for the foundation of passive houses in earthquake-prone regions. Eng Struct 112:133–145

Blocon (2015) Heat 2D/3D. Lund, Sweden

Ge H, Mcclung VR, Zhang S (2013) Impact of balcony thermal bridges on the overall thermal performance of multi-unit residential buildings: a case study. Energy Build 60:163–173

Goulouti K, De Castro J, Vassilopoulos AP, Keller T (2014) Thermal performance evaluation of fiber-reinforced polymer thermal breaks for balcony connections. Energy Build 70:365–371

Graphisoft (2015) ArchiCAD 19.0.0.—EcoDesigner STAR. Budapest, Hungary

IBO (2008) Details for passive houses: a catalogue of ecologically rated constructions, 3rd edn. Springer-Verlag, Vienna

Open Access This chapter is licensed under the terms of the Creative Commons Attribution 4.0 International License (http://creativecommons.org/licenses/by/4.0/), which permits use, sharing, adaptation, distribution and reproduction in any medium or format, as long as you give appropriate credit to the original author(s) and the source, provide a link to the Creative Commons license and indicate if changes were made.

The images or other third party material in this chapter are included in the chapter's Creative Commons license, unless indicated otherwise in a credit line to the material. If material is not included in the chapter's Creative Commons license and your intended use is not permitted by statutory regulation or exceeds the permitted use, you will need to obtain permission directly from the copyright holder.

Chapter 6
Conclusions

In recent years, energy savings have gained strategic importance due to requirements for less environmental pollution, the use of renewable energy sources, and ensuring energy independence (UN (United Nations) 2015). The latter was indicated by the adoption of measures to reduce the use of energy in the building sector, which constitutes a significant share of the total use of energy. Due to a lack of primary energy sources in Europe, most measures and details to increase energy efficiency in buildings originate from Northern and Central Europe with a cold continental climate, where energy savings for heating are great and the initial investment in better energy efficiency is repaid quickly. From the environmental and energy-efficiency aspect, solutions to reduce energy consumption also apply to other parts of Europe to attain the objectives of the European Union, e.g. the Directive on the energy performance of buildings (DEUS 2010/31/EU) and the European Green Deal (EU Commission 2019). Many existing solutions spread to other parts of the world. They include the transfer of details to provide energy efficiency to areas with increased seismic risk, where the established construction practice differs from non-earthquake-prone areas in many ways. Therefore, when using developed details and solutions to increase energy efficiency, it must be additionally verified in seismically active areas whether they can directly or indirectly affect the earthquake resistance of a building. Since there are no systematic and standard solutions for earthquake-prone areas, we studied the most common details used to reduce energy consumption in the monograph, which appear in practice in modern energy-efficient buildings.

The primary purpose of the monograph was to explore special structural details in energy-efficient buildings in the context of their earthquake resistance. The hypothesis that the principles of energy-efficient buildings (particularly requirements to prevent thermal bridges) can reduce the earthquake resistance of structures in comparison with conventional earthquake-resistant construction was put forward. In the first part of the monograph, we selected and presented structural details and solutions on the assumption that they are critical to earthquake resistance (Chap. 4). Various structural details on the building envelope for preventing thermal bridges and their interaction with the load-bearing structure were analysed, such as foundations on thermal insulation under the foundation slab, the building connection between an outer wall and the foundation slab, the building connection between the load-bearing balcony structure and an outer wall, the building connection between an outer wall and the unheated basement, the building connection between an outer wall and the roof structure, and others.

For the foundations on thermal insulation boards from extruded polystyrene (XPS), preliminary detailed numerical and experimental analyses of the seismic response were carried out (Azinović et al. 2014, 2015, 2016; Kilar et al. 2014). The results of the research show that foundations on thermal insulation under the foundation slab elongates the basic fundamental period of the structure. This is not always favourable. For instance, the stiff structures with short fundamental periods could be consequently shifted to the response spectrum plateau, where the seismic forces are larger. Many energy-efficient buildings would fit exactly in this category and therefore their shift to earthquake-prone areas could bring larger seismic forces and deformations. It was also established that foundations on thermal insulation may lead to the undesired rocking of a building on a flexible XPS layer and (uncontrolled) horizontal displacements at the connection between the foundation slab and thermal insulation or between individual layers of thermal insulation. The analyses results showed that rocking may result in exceeded elastic compressive deformation in XPS in three to four-storey slender and heavier buildings in areas with design ground acceleration higher than 0.25 g. In addition, such buildings (i.e. buildings higher than four storeys with a slender floor area and greater mass) may have instability issues as a result of rocking, which is why foundations on thermal insulation boards are not recommended in such cases. Greater carefulness and individual consideration are crucial to special cases of more complex buildings, irregular buildings in terms of floor plan or height and asymmetrical buildings, where the maximum number of storeys may be reduced. Such cases occur in buildings with large cantilevers or other height-related irregularities, and require the supervision of compressive stresses in thermal insulation under the foundation slab due to vertical and horizontal seismic forces. Based on the results of the analysed structures, we also find that lightweight buildings with less storeys are most exposed to the horizontal displacements on thermal insulation layer. The shift largely depends on the static friction coefficient in the selected foundation slab structural assembly and other boundary conditions (underground, non-underground building, etc.). In addition to the structural characteristics, the effect of insulation under the foundation slab on the use of energy and the prevention of thermal bridges was also evaluated.

Solutions including base insulation blocks for masonry structures in earthquake-prone areas were also analysed. Based on the literature review and the characteristics of base insulation blocks used as masonry in non-earthquake-prone areas, we find that most of them are inappropriate due to the insufficient normalised compressive strength. The limitation to the minimum compressive strength of masonry is provided in standards, such as Eurocode 8, and is intended to prevent undesired masonry failure mechanisms in earthquake-prone areas. We also analysed the effect of the thermal conductivity of base insulation blocks on the extent of the thermal bridge and surface temperatures of the building connection detail with the unheated basement. The effect of load-bearing thermal insulation elements on resolving the thermal bridge at the location where cantilever structures are fixed, was also analysed and demonstrated. All the performed analyses and reviewed literature support the importance of energy-efficient details, reflected in higher surface temperatures on the inner surface of the envelope and lower heat losses.

For the detail of cantilever structures with load-bearing thermal insulation elements, preliminary detailed numerical and experimental analyses of the seismic response were carried out (Azinović et al. 2015). It was established that the analysed load-bearing thermal insulation elements could be used for cantilevers of up to 300 cm in length in earthquake-prone areas on the basis of limit deflection ($w < l/150$) and without additional measures. The results of the analysis of these elements also confirmed the hypothesis that the cantilever uplift and tensile stress in the bottom edge of the cross-section could occur in an earthquake. Since most load-bearing thermal insulation elements on the market originate from non-earthquake-prone areas, such precast elements are designed exclusively for vertical loads with an installed asymmetrical steel reinforcement (only in the top edge of the cross-section). Based on the analyses, we believe that the elements must be improved for earthquake-prone areas (e.g. by installing reinforcement also in the bottom edge of the cross-section or other measures) to avoid potential severe damage to them resulting from the cantilever uplift in severe earthquakes.

In addition to the shown importance of using modern energy-efficient details, the monograph also focuses on the effect of these solutions on the earthquake resistance of buildings. The assumptions of their effect on earthquake resistance were presented for all the analysed details. Preliminary studies show that earthquake resistance is most significantly affected by the requirement of the continuous thermal envelope, which is a condition necessary to prevent thermal bridges and their adverse consequences. Therefore, we proposed the detail evaluation methodology based on guidelines for energy-efficient and earthquake-resistant construction. Two parts of evaluation are defined, whereby one part is about the quality of a detail from the technical and structural aspect and the other from the environmental and energy-efficiency aspect. Each detail must be assessed on the basis of seven criteria in the environmental and energy-efficiency evaluation, and seven criteria in the technical and structural evaluation. The final score may also be affected by weighting factors and six external factors, which affect evaluation less than the assessment of the primary criteria. All the parameters and the prepared evaluation criteria are described in detail, and guidelines for the design of structural energy-efficient buildings in earthquake-prone areas are provided. The evaluation methodology facilitates the separation of details that are more critical from the aspect of earthquake resistance and energy efficiency on the basis of simple engineering approaches (the review of designs and geometry, analyses of heat transfer, etc.). On the other hand, individual parameters can be assessed on the basis of detailed experimental analyses and (or) numerical simulation to precisely determine the quality of a detail in terms of its structural resistance (with a focus on seismic loads), energy efficiency, and environmental protection.

In the last section of the monograph, the proposed evaluation methodology was used in practical cases of four different types of details: the building connection between an outer wall and the foundation slab, the building connection between an outer wall and the unheated basement, the building connection between the load-bearing balcony structure and an outer wall, and the building connection between an outer wall and the roof. The assessment results of analysed details showed that the proposed methodology can be used to separate better solutions from poorer ones in the conceptual design. In this way, significant changes to the load-bearing capacity, stiffness, interruptions in the load-bearing structure, asymmetrical solutions, and other important characteristics that affect the quality of a detail to be used in earthquake-resistant structures are recognised. Analyses of heat transfer and environmental impact scores can also be used to choose from alternative detail solutions and decide, on the basis of the evaluation results, on the measures to prevent thermal bridges.

It can be concluded that the proposed methodology is generally not intended for the absolute evaluation of a selected detail, but rather for a comparison of details and to assist designers in the conceptual phase of designing the building envelope, serving as a tool for the selection of the best solutions. The limit values of the evaluation criteria could be determined by users in view of the conditions of the location and local regulations (rules and criteria applicable to the construction of energy-efficient and earthquake-resistance buildings in the analysed area).

The appendix to the monograph includes a catalogue of the selected structural details most frequently used in energy-efficient buildings, focusing on both problematic details as well as details that contribute to better solutions in practice. Various unsuitable solutions used to appear, in which energy efficiency was not crucial, resulting in energy-wasting buildings and, as experience shows, other adverse effects on thermal comfort and health of users of the building (e.g. consequences of thermal bridges—condensation and mould). In our opinion, one of the advanced options is also to use the methodology in computer programmes used for building information modelling (BIM). Some of these programmes already facilitate the analysis of thermal bridges, the environmental analysis, and the analysis of the energy use, which is required in the environmental and energy-efficiency evaluation. The latter could be upgraded with the technical and structural evaluation of the most common details, facilitating the selection of the most appropriate details for earthquake-prone areas during design.

The monograph is a result of over ten years of research by the authors in the field of the earthquake resistance of modern energy-efficient buildings. It aims to promote a wider interest in, and awareness of, the importance of structural details and their effect on the structural safety of energy-efficient buildings. At the same time, the monograph is the basis for further in-depth and interdisciplinary studies in this field. The cooperation of all professions included in the design of energy-efficient buildings and their end users is crucial.

References

Azinović B, Koren D, Kilar V (2014) The seismic response of low-energy buildings founded on a thermal insulation layer—a parametric study. Eng Struct 81:398–411

Azinović B, Kilar V, Koren D (2015) Erdbebensicherheit vorgefertigter wärmegedämmter Stahlbeton-Konsolenelemente. Bauingenieur 90:489–499

Azinović B, Kilar V, Koren D (2016) Energy-efficient solution for the foundation of passive houses in earthquake-prone regions. Eng Struct 112:133–145

Commission EU (2019) The European green deal, Belgium, Brussels

DEUS (2010/31/EU) Direktiva o energetski učinkovitosti stavb (Online). Available: http://eur-lex.europa.eu/LexUriServ/LexUriServ.do?uri=OJ:L:2010:153:0013:0035:SL:PDF. Accessed 12 July 2013

Kilar V, Koren D, Bokan-Bosiljkov V (2014) Evaluation of the performance of extruded polystyrene boards—implications for their application in earthquake engineering. Polym Test 40:234–244

Kilar V, Azinović B, Koren D (2016) Foundation concept for passive houses in seismic areas (in Slovene). Gradbeni Vestnik 65:59–70

UN (United Nations) (2015) UN sustainable development goals (Online). Available: https://sustainabledevelopment.un.org/sdgs. Accessed 15 Jan 2021

Open Access This chapter is licensed under the terms of the Creative Commons Attribution 4.0 International License (http://creativecommons.org/licenses/by/4.0/), which permits use, sharing, adaptation, distribution and reproduction in any medium or format, as long as you give appropriate credit to the original author(s) and the source, provide a link to the Creative Commons license and indicate if changes were made.

The images or other third party material in this chapter are included in the chapter's Creative Commons license, unless indicated otherwise in a credit line to the material. If material is not included in the chapter's Creative Commons license and your intended use is not permitted by statutory regulation or exceeds the permitted use, you will need to obtain permission directly from the copyright holder.

Appendix
Examples of the Use of the Methodology for Evaluating Structural Details

The appendix includes additional examples of the evaluation of structural details according to the proposed methodology. Input data to determine environmental and energy-efficiency parameters are the same as in Chap. 5. An analysis of heat transfer for each structural detail was carried out in the *Archicad* programme, based on which the temperature range and extent of the thermal bridge were determined by calculating linear thermal transmittance. To determine the technical and structural parameters, a simplified engineering approach was used, which is based on the comparative evaluation according to existing professional and scientific literature, and the analyses performed in Azinović et al. (2015), Azinović et al. (2016), Azinović et al. (2014), Kilar et al. (2014). External and weighting factors were not taken into account in the performed analyses whose results are shown in this appendix (neutral vales were used - $\gamma_{i,Ki} = \gamma_{i,Ei} = 1$).

The evaluation results for each detail are shown in a special form on two pages. The first page includes a brief description, the characteristics of a detail, the composition of structural assemblies, and the heat transfer simulation results (building physics). On the second page, the results are shown in a table, and radial and column diagrams. Concluding remarks are given at the end of the form, which include a brief explanation of the final evaluation score for each detail, and potential flaws of the detail.

Similar to Chap. 5, the evaluation results are shown for various types of details:

- Building connection between an outer wall and the foundation slab (Designation: *DT-foundation detail*);
- Building connection between an outer wall and the unheated basement (Designation: *DK-basement detail*);
- Building connection between the load-bearing balcony structure and an outer wall (Designation: *DB-balcony detail*);
- Building connection between an outer wall and the roof (Designation: *DS-roof detail*).

Details are divided into groups by their types. The evaluation results of various types of details are not directly comparable, as external and weighting factors are not taken into account. For example, the final average scores for details from the *DB* and *DK* groups are not comparable, as their comparison would be misleading. Only scores for details within a group (e.g. *DK*-01 and *DK*-02) could be compared to a certain extent. However, even in this case, caution must be exercised when drawing conclusions on the basis of comparisons, since certain results of the technical and structural evaluation are determined conservatively because they are not based on experimental tests or accurate numerical analyses.

Most shown solutions for details appropriate for passive houses are summarised from (IBO 2008) or the catalogues of manufacturers of prefabricated solutions to prevent thermal bridges (e.g. Schöck (2020), Wienerberger (2016)). Certain details were also selected as reference examples of poor practice (e.g. detail *DB*-01: detail of the uninsulated balcony slab). IBO (2008) have performed the LCA analysis of all analysed structural assemblies for a catalogue, which we used to determine the environmental and energy-efficiency evaluation according to the proposed methodology.

Detail code	*DT*-01
Brief description	Building connection detail between an RC outer wall and strip foundations

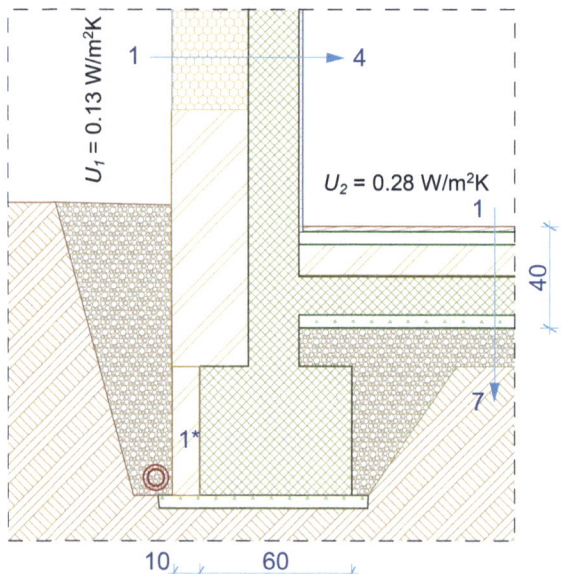

Appendix: Examples of the Use of the Methodology for Evaluating Structural Details 143

Detail Properties

The connection between an RC outer wall and the strip foundations and the RC slab on ground is analysed. Generally, a thermal bridge occurs towards the strip foundations. Also analysed is the slab on ground structural assembly with high thermal transmittance, which does not comply with the requirements of the PH standard. Thermal bridges through the strip foundations are reduced by extending TI (e.g. XPS) to the bottom of the strip foundations (1*).

The detail is composed of:

- RC outer wall insulated with EPS ($U_1 = 0.13$ W/(m² K))
- RC slab on ground ($U_2 = 0.28$ W/(m² K)).

Structural Assemblies

Outer wall Thermal transmittance $U_1 = 0.13$ W/(m² K)			Slab on ground Thermal transmittance $U_2 = 0.28$ W/(m² K)		
No.	Material	T (cm)	No.	Material	T (cm)
1	Silicate thin-layer plastering	–	1	Flooring–wooden parquet	1.5
2	EPS (expanded polystyrene)	30.0	2	Concrete screed	5.0
3	RC outer wall	20.0	3	XPS (extruded polystyrene)	12.0
4	Finishes	–	4	Bitumen cardboard as waterproofing in two layers	1.0
			5	RC slab	15.0
			6	Blinding concrete	5.0
			7	Drainage layer	15.0

144 Appendix: Examples of the Use of the Methodology for Evaluating Structural Details

Building Physics

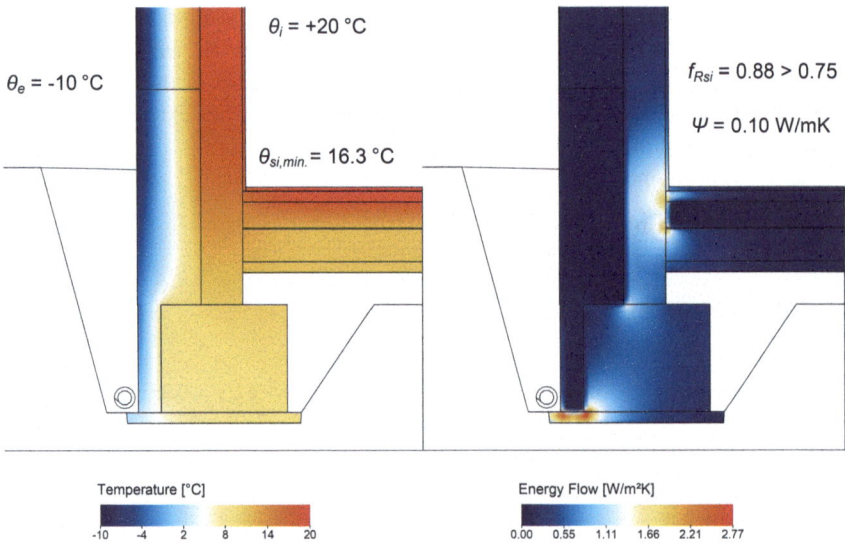

Detail Evaluation

		Score	Selected weighting factors (weights)	Corrected weighting factors (influence of external factors)	Share of weighted score
Environmental & energy-efficiency parameters	E1	3	1	1	0.43
	E2	2	1	1	0.29
	E3	3	1	1	0.43
	E4	4	1	1	0.57
	E5	6	1	1	0.86
	E6	3	1	1	0.43
	E7	5	1	1	0.71
Technical & structural parameters	K1	6	1	1	0.86
	K2	5	1	1	0.71
	K3	5	1	1	0.71
	K4	5	1	1	0.71
	K5	6	1	1	0.86
	K6	6	1	1	0.86
	K7	5	1	1	0.71

Appendix: Examples of the Use of the Methodology for Evaluating Structural Details 145

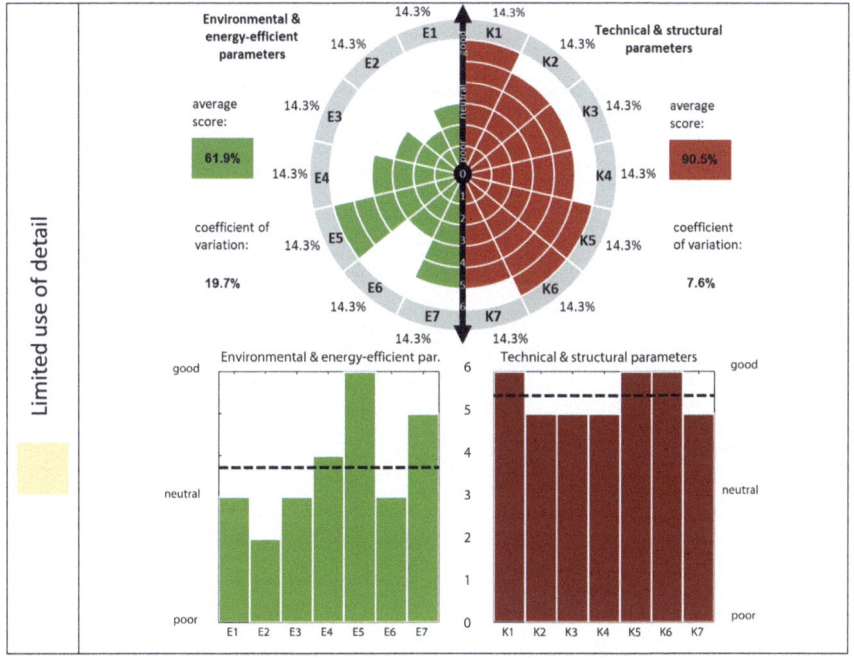

Concluding Remarks

Generally, the environmental and energy-efficiency score of the strip foundation details is lower, since it is difficult to avoid thermal bridges through the strip foundations. High concentrations of compressive stresses occur below these foundations, which usually exceed the compressive strength of the most common thermal insulation materials even in lower buildings. Extending TI to the bottom of the strip foundations prevents condensation and mould, but cannot ensure complete thermal comfort. The situation is completely different from the technical and structural aspect, from which the detail's score is very good. This detail includes a continuous transfer of vertical loads to the foundation base. There was a slight deduction on the account of soil deformability, which can significantly affect the load-bearing capacity (resistance) of the whole foundation detail. The score due to poor/good strength of the soil can be further decreased or increased with external parameter Z1-building location.

Detail code	*DT*-02
Short description	Building connection detail between an outer masonry wall and an RC foundation slab

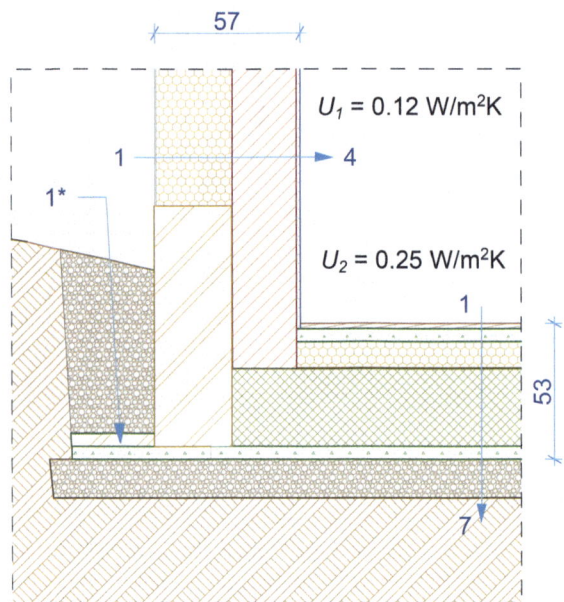

Detail Properties

Detail *DT*-02 is a solution for the contact between an outer masonry wall and an RC foundation slab thermally insulated on the internal side. The analysed slab on ground structural assembly has high thermal transmittance, which does not comply with the requirements of the PH standard, but is admissible by some European technical guidelines ($U < 0.30$ W/(m² K)). Also, a thermal bridge cannot be prevented without using an insulation base block. In practice, such solutions are more common in buildings with a basement. 1* is additional thermal insulation against the freezing of the ground

The detail is composed of:

- outer masonry wall insulated with EPS ($U_1 = 0.12$ W/(m² K))
- RC slab on ground ($U_2 = 0.25$ W/(m² K)).

Structural Assemblies

Outer wall Thermal transmittance $U_1 = 0.12$ W/(m² K)			Slab on ground Thermal transmittance $U_2 = 0.25$ W/(m² K)		
No.	Material	T (cm)	No.	Material	T (cm)
1	Silicate thin-layer plastering	–	1	Flooring–wooden parquet	1.5
2	EPS (expanded polystyrene)	30.0	2	Concrete screed	5.0
3	RC outer wall	25.0	3	EPS (expanded polystyrene)	10.0
4	Cement mortar	1.5	4	Bitumen cardboard as waterproofing in two layers	1.0
			5	RC slab	30.0
			6	Blinding concrete	5.0
			7	Drainage layer	15.0

Building Physics

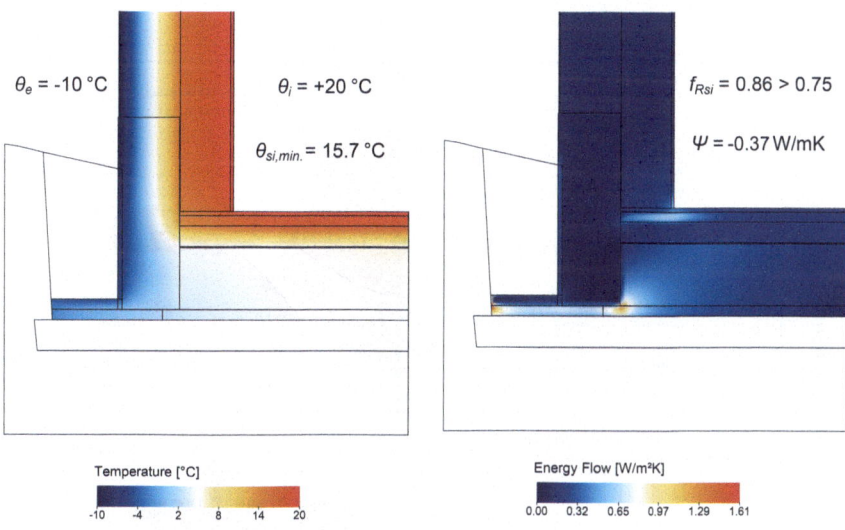

$\theta_e = -10\ °C$
$\theta_i = +20\ °C$
$\theta_{si,min.} = 15.7\ °C$
$f_{Rsi} = 0.86 > 0.75$
$\Psi = -0.37$ W/mK

Temperature [°C]
-10 -4 2 8 14 20

Energy Flow [W/m²K]
0.00 0.32 0.65 0.97 1.29 1.61

Detail Evaluation

		Score	Selected weighting factors (weights)	Corrected weighting factors (influence of external factors)	Share of weighted score
Environmental & energy-efficiency parameters	E1	3	1	1	0.43
	E2	2	1	1	0.29
	E3	3	1	1	0.43
	E4	4	1	1	0.57
	E5	6	1	1	0.86
	E6	3	1	1	0.43
	E7	5	1	1	0.71
Technical & structural parameters	K1	5	1	1	0.71
	K2	5	1	1	0.71
	K3	4	1	1	0.57
	K4	5	1	1	0.71
	K5	6	1	1	0.86
	K6	5	1	1	0.71
	K7	5	1	1	0.71

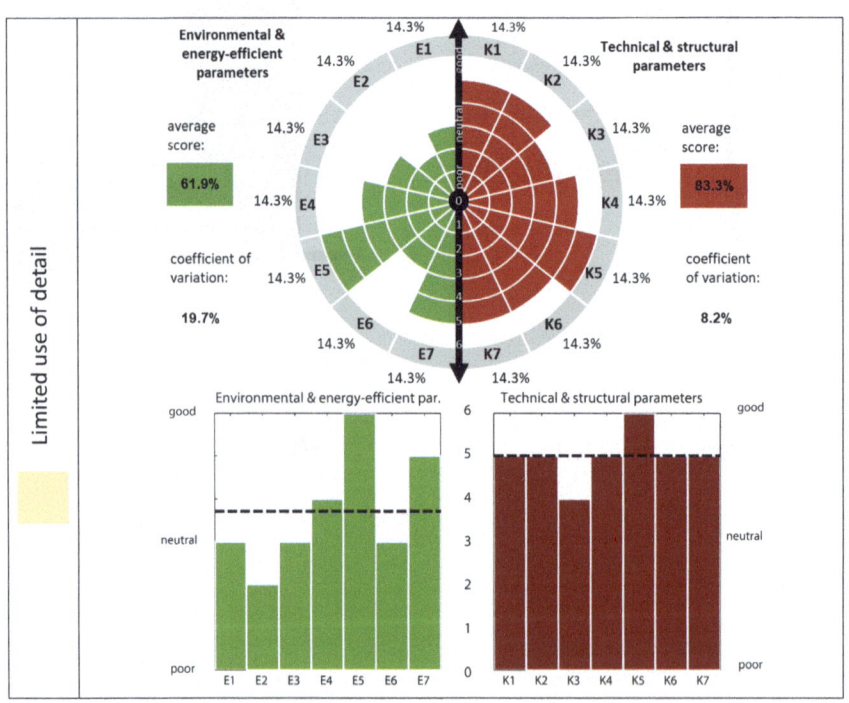

Concluding Remarks

From the aspect of the environmental and energy-efficiency parameters, the described foundation slab detail with minimum TI thickness on the internal side is admissible but undesired, since the effect of a geometrical thermal bridge must be taken into account in the analysis of the use of energy in the building. Also surface temperature is reduced to approx. 15–16 °C, which is insufficient from the aspect of thermal comfort. Negative effects can be further increased/reduced with external parameters Z2 (importance of the building) and Z3 (effect on the use of energy in the building). The detail score from the aspect of structural safety is very good. Deduction from the highest score is on account of the flexibility of the soil and filling/underground section, which can affect the global seismic safety of the building. This effect can be further reduced or increased with external parameter Z1-building location.

Detail code	*DT*-03
Brief description	**Building connection detail between a cross-laminated timber (CLT) wall and an RC foundation slab**

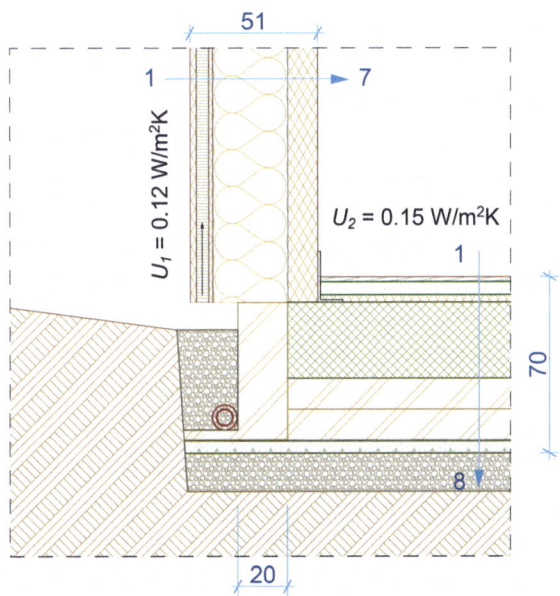

Detail Properties

Detail DT-03 is a solution for the contact between an outer CLT wall and an RC foundation slab thermally insulated at the contact with the ground. From the aspect of thermal transmittance, both structural assemblies comply with the PH standard, and thermal envelope is continuous, preventing a thermal bridge through the foundation slab. The CLT wall is unilaterally fixed to the RC slab with L-shaped steel angle brackets.

The detail is composed of:

- ventilated timber wall insulated with mineral wool ($U_1 = 0.12$ W/(m^2 K))
- RC foundation slab on TI boards from XPS ($U_2 = 0.15$ W/(m^2 K)).

Structural Assemblies

Outer wall Thermal transmittance $U_1 = 0.12$ W/(m^2 K)			Slab on ground Thermal transmittance $U_2 = 0.15$ W/(m^2 K)		
No.	Material	T (cm)	No.	Material	T (cm)
1	Wood panelling	2.5	1	Flooring–wooden parquet	1.5
2	Ventilated layer	5.0	2	Concrete screed	5.0
3	MDF	1.6	3	Acoustic mineral wool	3.0
4	Mineral wool in timber substructure	30.0	4	RC foundation slab	30.0
5	PE vapour barrier	–	5	XPS boards (2 × 12 cm)	24.0
6	Wall from cross-laminated timber panels	12.0	6	Bitumen cardboard as waterproofing in two layers	1.0
7	Finishes	–	7	Blinding concrete	5.0
			8	Drainage layer	15.0

Appendix: Examples of the Use of the Methodology for Evaluating Structural Details

Building Physics

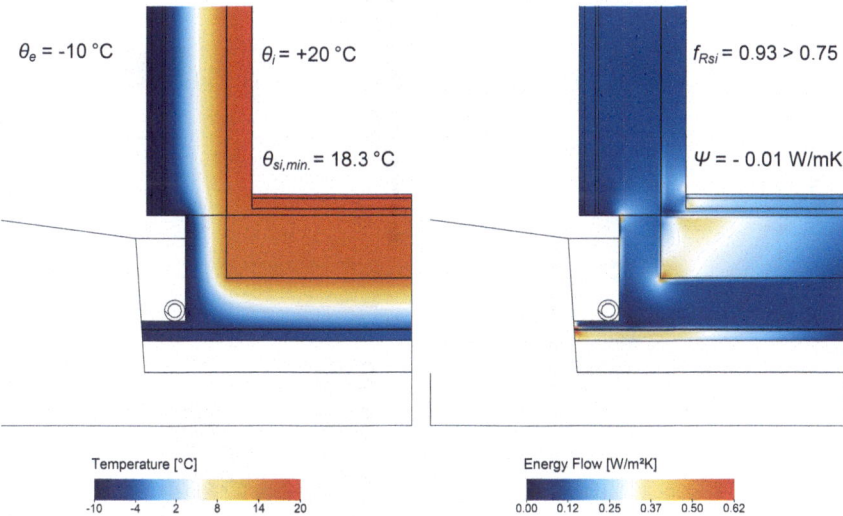

$\theta_e = -10\ °C$ $\theta_i = +20\ °C$ $f_{Rsi} = 0.93 > 0.75$

$\theta_{si,min.} = 18.3\ °C$ $\Psi = -0.01\ W/mK$

Temperature [°C]
Energy Flow [W/m²K]

Detail Evaluation

		Score	Selected weighting factors (weights)	Corrected weighting factors (influence of external factors)	Share of weighted score
Environmental & energy-efficiency parameters	E1	5	1	1	0.71
	E2	5	1	1	0.71
	E3	6	1	1	0.86
	E4	6	1	1	0.86
	E5	3	1	1	0.43
	E6	4	1	1	0.57
	E7	2	1	1	0.29
Technical & structural parameters	K1	4	1	1	0.57
	K2	4	1	1	0.57
	K3	1	1	1	0.14
	K4	1	1	1	0.14
	K5	2	1	1	0.29
	K6	3	1	1	0.43
	K7	3	1	1	0.43

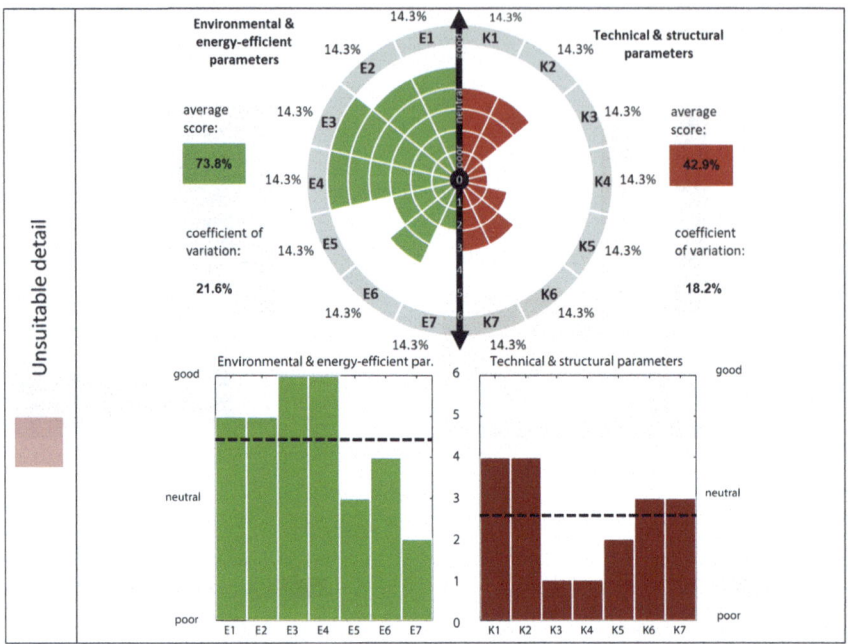

Concluding Remarks

The score of detail *DT*-03 is lower from the technical and structural aspect due to lower parameters resulting from inappropriate connections of the CLT wall with foundations on thermal insulation. Fixing the CLT wall with L-shaped shear steel angle brackets was taken into consideration, which is suitable in low seismicity areas, but is insufficient in earthquake-prone areas, where also hold-down connectors should be used due to rocking wall movement. Therefore, a reduction in K3 was applied, since more damage can be expected as a result from the asymmetrical/inappropriate use of connections, which makes the wall less ductile and worsens the response in the event of an earthquake. There is also a higher probability that sliding will occur between thermal insulation boards in light timber frame structures than in mass timber structures. From the environmental and energy-efficiency aspect, the detail's score is good on account of its continuous thermal insulation, prevented thermal bridges, and low thermal transmittance of the selected structural assemblies.

Appendix: Examples of the Use of the Methodology for Evaluating Structural Details 153

Detail code	*DT-04*
Brief description	Building connection detail between a masonry wall and an RC foundation slab on TI

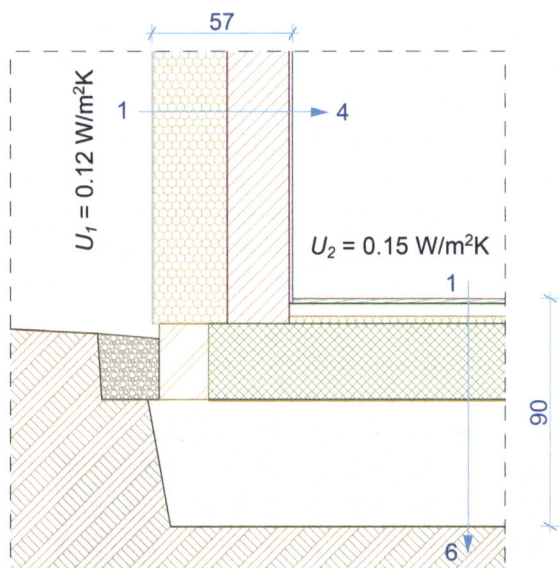

Detail Properties

The special feature of detail DT-04 is its foundation on insulation aggregate, which is most frequently from cellular glass. This measure improves the thermal transmittance of the structural assembly made of the foundation slab on ground, and eliminates a thermal bridge at the contact between the foundation slab and an outer wall. The load-bearing capacity of the insulation aggregate depends on the basic material (e.g. cellular glass) and the preparation of the ground. Increased thermal transmittance in areas with groundwater must also be taken into account.

The detail is composed of:

- Outer masonry wall insulated with EPS ($U_1 = 0.12$ W/(m^2 K))
- RC foundation slab on insulation ($U_2 = 0.15$ W/(m^2 K)).

Structural Assemblies

Outer wall Thermal transmittance $U_1 = 0.12$ W/(m² K)			Slab on ground Thermal transmittance $U_2 = 0.15$ W/(m² K)		
No.	Material	T (cm)	No.	Material	T (cm)
1	Silicate thin-layer plastering	–	1	Flooring–wooden parquet	1.5
2	EPS (expanded polystyrene)	30.0	2	Concrete screed	5.0
3	Outer masonry wall	25.0	3	Acoustic mineral wool	3.0
4	Cement mortar	1.5	4	Waterproof RC foundation slab	30.0
			5	Cellular glass gravel insulation	50.0
			6	PP geotextile	–

Building Physics

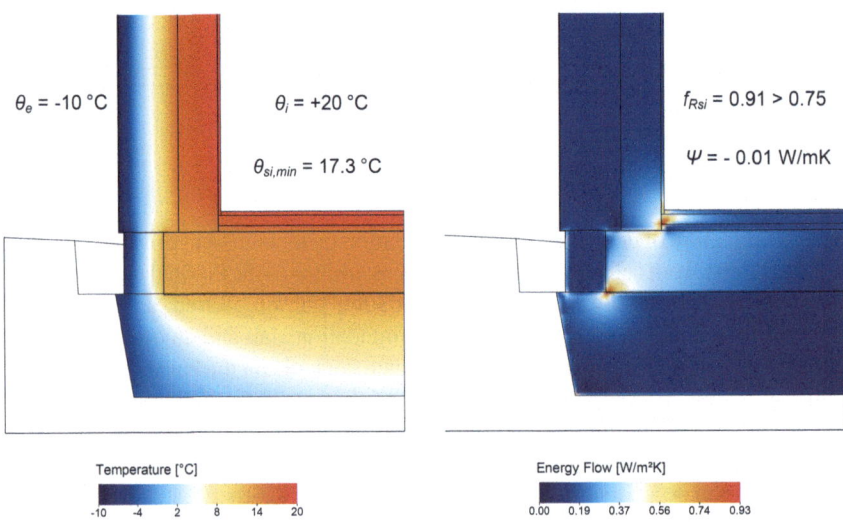

$\theta_e = -10\ °C$ $\theta_i = +20\ °C$

$\theta_{si,min} = 17.3\ °C$

$f_{Rsi} = 0.91 > 0.75$

$\Psi = -0.01$ W/mK

Temperature [°C]
-10 -4 2 8 14 20

Energy Flow [W/m²K]
0.00 0.19 0.37 0.56 0.74 0.93

Appendix: Examples of the Use of the Methodology for Evaluating Structural Details 155

Detail Evaluation

		Score	Selected weighting factors (weights)	Corrected weighting factors (influence of external factors)	Share of weighted score
Environmental & energy-efficiency parameters	E1	5	1	1	0.71
	E2	5	1	1	0.71
	E3	5	1	1	0.71
	E4	6	1	1	0.86
	E5	6	1	1	0.86
	E6	1	1	1	0.14
	E7	3	1	1	0.43
Technical & structural parameters	K1	5	1	1	0.71
	K2	5	1	1	0.71
	K3	3	1	1	0.43
	K4	4	1	1	0.57
	K5	5	1	1	0.71
	K6	4	1	1	0.57
	K7	4	1	1	0.57

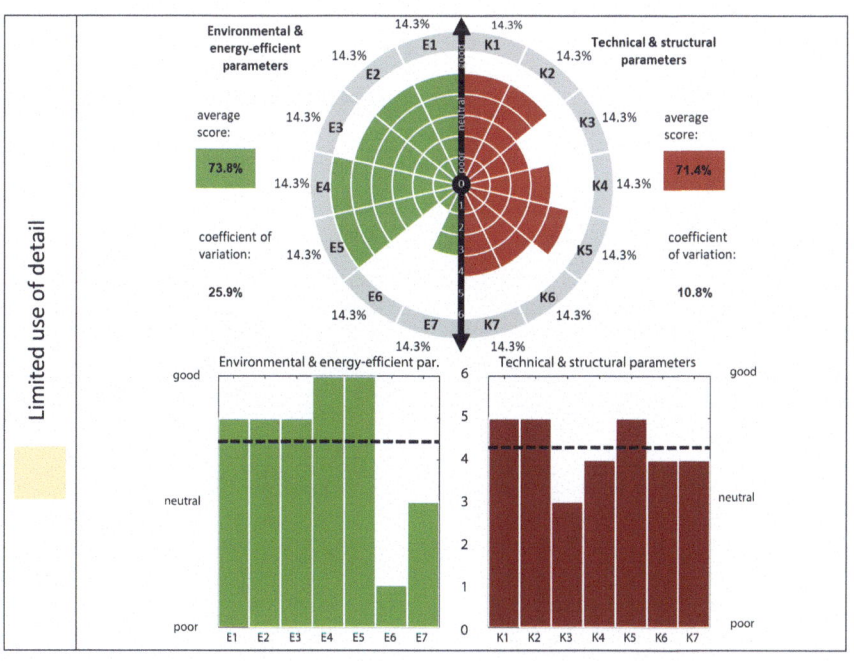

Concluding Remarks

Detail *DT-04* has positive environmental and energy-efficiency, and technical and structural scores. The detail with insulation aggregate can also be used in earthquake-prone areas if the ground is well prepared and the material has a good load-bearing capacity. The score can be decreased/increased with external parameter Z4 (complexity of construction, possibility of contractor failure). From the environmental and energy-efficiency aspect, deduction is made on account of the durability and functioning of the structural assembly in the event of high groundwater, since in this case the aggregate can lose its insulation properties. The quantity of material must also be taken into account, since the thermal conductivity of insulation aggregate from cellular glass is higher, which requires greater total thickness of the structural assembly to achieve the same thermal transmittance as in structural assemblies from thermal insulation boards (e.g. XPS boards). The latter can affect the final price, which may be taken into account with external parameter Z6.

Detail code	*DT*-05
Brief description	**Building connection detail between an outer masonry wall and an RC foundation slab**

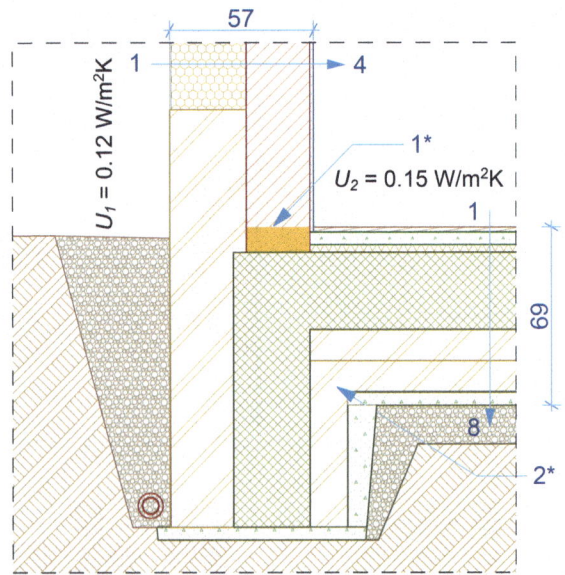

Appendix: Examples of the Use of the Methodology for Evaluating Structural Details 157

Detail Properties

Detail DT-05 is a building connection between an outer masonry wall and the RC foundation slab thermally insulated with TI boards (XPS). At the location where the outer wall is fixed, the foundation slab is reinforced with a foundation beam. A higher heat flow towards the foundation beam must be taken into account, which is limited with perlite-filled bricks in the first row of masonry units (1*). The foundation beam is thermally insulated on both sides (2*).

The detail is composed of:

- Outer masonry wall insulated with EPS ($U_1 = 0.12$ W/(m² K))
- RC foundation slab on TI boards from XPS ($U_2 = 0.15$ W/(m² K)).

Structural Assemblies

Outer wall Thermal transmittance $U_1 = 0.12$ W/(m² K)			Slab on ground Thermal transmittance $U_2 = 0.15$ W/(m² K)		
No.	Material	T (cm)	No.	Material	T (cm)
1	Silicate thin-layer plastering	–	1	Flooring–wooden parquet	1.5
2	EPS (expanded polystyrene)	30.0	2	Concrete screed	5.0
3	Masonry outer wall	25.0	3	Acoustic mineral wool	3.0
4	Cement mortar	1.5	4	RC foundation slab	30.0
			5	XPS boards (2 × 12 cm)	24.0
			6	Bitumen cardboard as waterproofing in two layers	1.0
			7	Blinding concrete	5.0
			8	Drainage layer	15.0

Building Physics

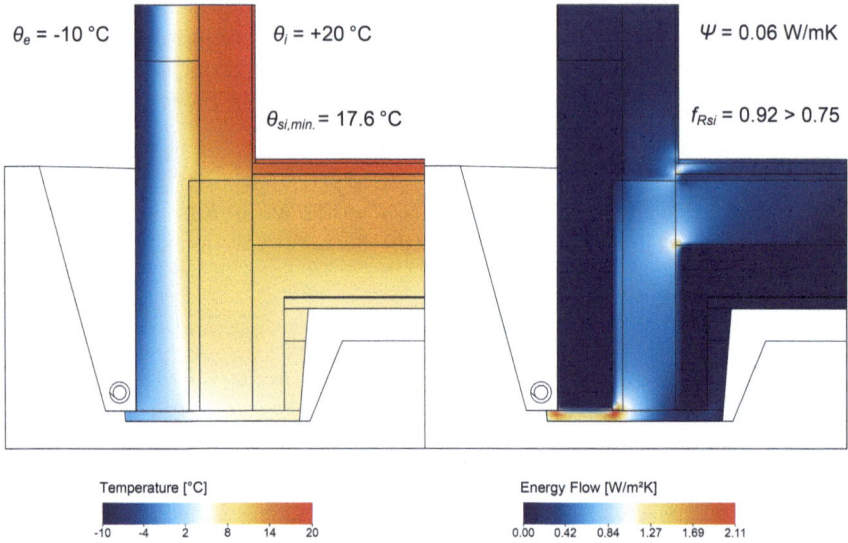

Detail Evaluation

		Score	Selected weighting factors (weights)	Corrected weighting factors (influence of external factors)	Share of weighted score
Environmental & energy-efficiency parameters	E1	6	1	1	0.86
	E2	3	1	1	0.43
	E3	5	1	1	0.71
	E4	5	1	1	0.71
	E5	6	1	1	0.86
	E6	2	1	1	0.29
	E7	5	1	1	0.71
Technical & structural parameters	K1	5	1	1	0.71
	K2	5	1	1	0.71
	K3	4	1	1	0.57
	K4	4	1	1	0.57
	K5	6	1	1	0.86
	K6	5	1	1	0.71
	K7	4	1	1	0.57

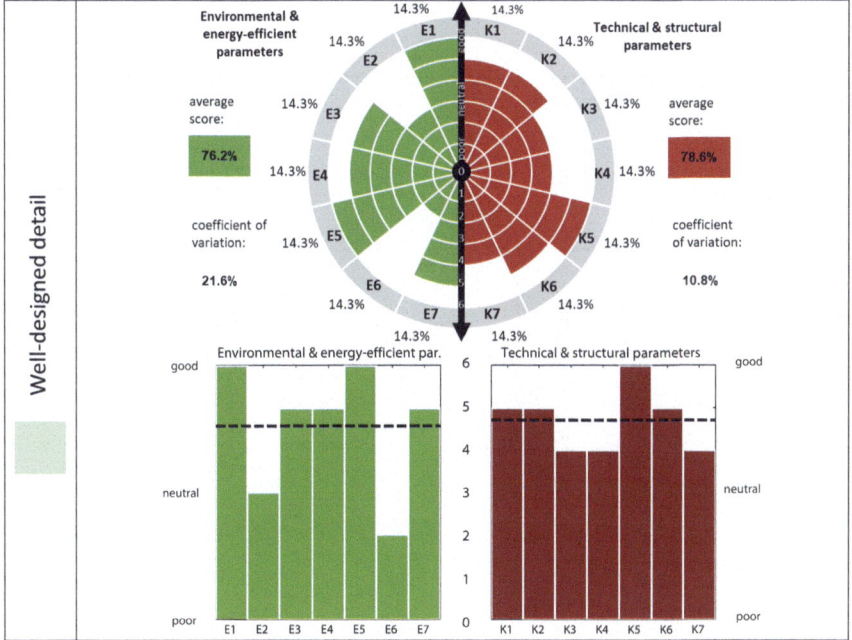

Concluding Remarks

The analysis includes a detail in which a thermal bridge is not completely prevented, but perlite-filled bricks are used to achieve better insulation at the location where the outer wall is fixed. The foundation beam is thermally insulated on both sides, which improves the environmental and energy-efficiency assessment of the detail. From the technical and structural aspect, perlite-filled bricks used as base insulation blocks have poorer compressive strength and other strength-related parameters than conventional modular bricks. On the other hand, the load in the analysed detail is transferred more directly through the foundation beam to the ground, giving the detail a slightly better structural score than the detail of the foundation slab on thermal insulation (see detail evaluation in Sect. 5.1).

Detail code	*DB*-01
Brief description	Building connection detail between an RC balcony slab and an outer masonry wall

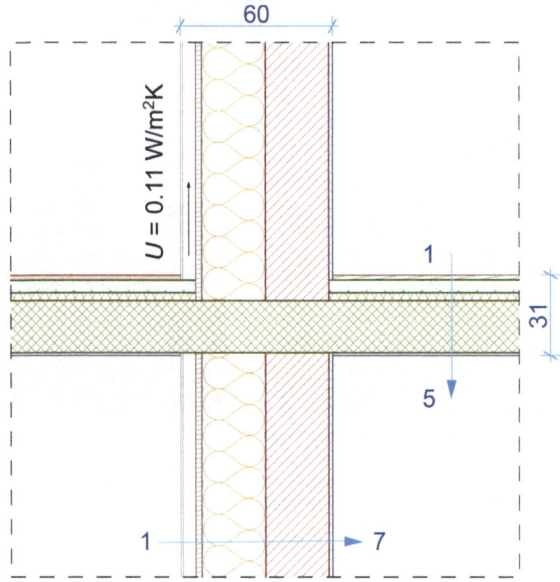

Detail Properties

The building connection detail between an RC balcony slab and an outer masonry wall with a ventilated wooden façade was selected for analysis. The outer wall structural assembly complies with the criteria of the PH standard regarding thermal transmittance, while the thermal bridge through the RC balcony slab was not eliminated. Such details must be avoided when constructing new buildings, but can occur in the restoration of the building envelope by increasing the thickness of TI. Increased thickness of TI can make thermal bridges even more noticeable than before restoration.

The detail is composed of:

- outer masonry wall insulated with min. wool ($U = 0.11$ W/(m^2 K))
- RC interstorey and balcony slab.

Appendix: Examples of the Use of the Methodology for Evaluating Structural Details 161

Structural Assemblies

Outer wall Thermal transmittance $U_1 = 0.11$ W/(m² K)			RC interstorey slab		
No.	Material	T (cm)	No.	Material	T (cm)
1	Fibre cement board	0.8	1	Flooring–wooden parquet	1.5
2	Ventilated layer	5.0	2	Concrete screed	5.0
3	Wind barrier (felt, geotextile)	0.1	3	Acoustic mineral wool	3.0
4	Plywood	2.4	4	RC load-bearing slab	20.0
5	Stone wool in timber substructure	25.0	5	Finishes	–
6	Hollow brick (Poroblock 29/25)	25.0			
7	Cement mortar	1.5			

Building Physics

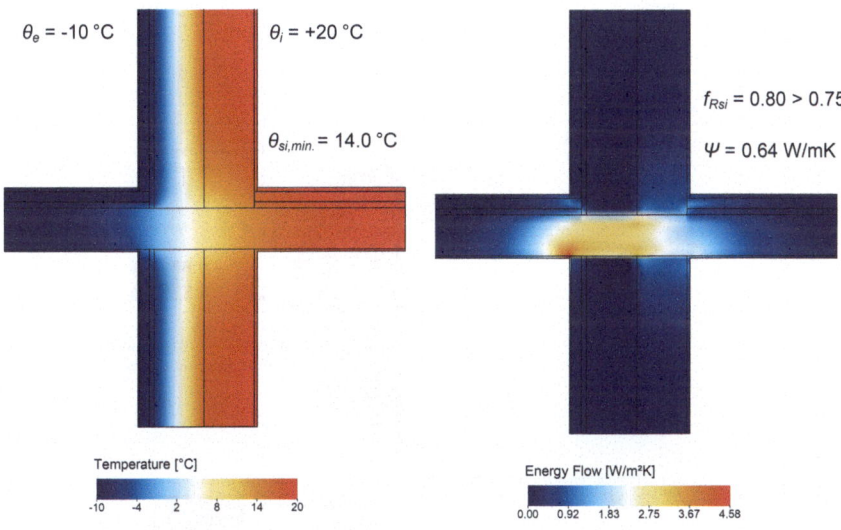

162 Appendix: Examples of the Use of the Methodology for Evaluating Structural Details

Detail Evaluation

		Score	Selected weighting factors (weights)	Corrected weighting factors (influence of external factors)	Share of weighted score
Environmental & energy-efficiency parameters	E1	6	1	1	0.86
	E2	0	1	1	0.00
	E3	1	1	1	0.14
	E4	0	1	1	0.00
	E5	2	1	1	0.29
	E6	2	1	1	0.29
	E7	0	1	1	0.00
Technical & structural parameters	K1	6	1	1	0.86
	K2	6	1	1	0.86
	K3	5	1	1	0.71
	K4	5	1	1	0.71
	K5	6	1	1	0.86
	K6	6	1	1	0.86
	K7	6	1	1	0.86

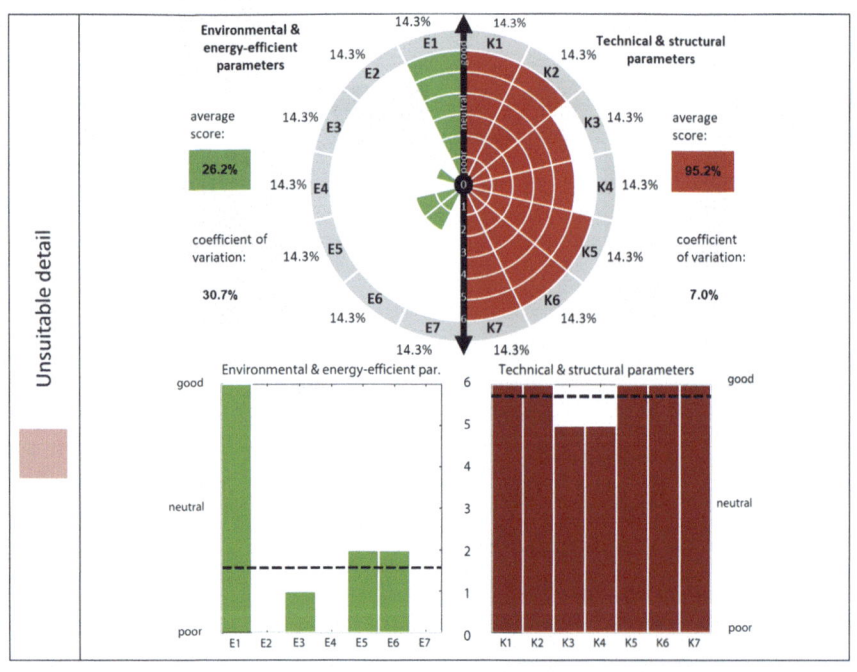

Concluding Remarks

In energy-efficient buildings, detail *DB*-01 must be replaced due to mould, poor thermal comfort, and excessive energy consumption for heating. It is only shown for comparison, since it received almost full score for technical and structural parameters but very few for environmental and energy-efficiency parameters. It is characteristic of the detail that its load-bearing structure has no special interventions (continued load-bearing structure, no interruptions with TI). Such a detail can be seen in restorations, where a decision must be made to replace the load-bearing structure with solutions to reduce or eliminate the thermal bridge. Increasing TI on the outer wall structural assembly (thermal transmittance as required by the PH standard) also increases the temperature on the inner surface of the detail ($\theta_{si,\min}$). In the analysed example, the temperature is, therefore, higher than limit temperature for condensation (mould) to occur, but the desired level of thermal comfort is not achieved.

Detail code	*DB*-02
Brief description	Building connection detail of a steel load-bearing structure and steel balcony cantilever beams

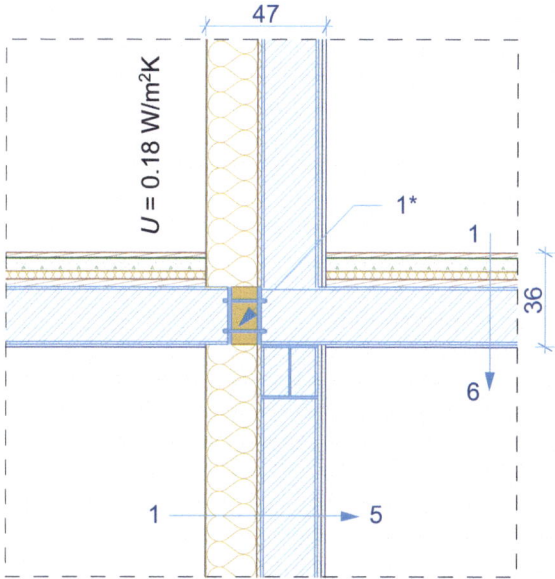

Detail Properties

A steel load-bearing structure with TI on the external side of the wall is analysed. In such a solution, a thermal bridge occurs at the location of the contact of the steel cantilever beam for the balcony structure (point thermal bridge). The thermal bridge is eliminated with a precast load-bearing TI element designed specifically for steel structures (1*). The load-bearing TI element is 8 cm thick and its thermal conductivity is $\lambda = 0.12$ W/(m K). The analysed element provides compressive support only in the lower part of the contact, while the upper part only contains bolts under tensile stress.

The detail is composed of:

- outer wall from steel beams insulated with min. wool (U = 0.18 W/(m² K))
- interstorey structure from steel beams.

Structural Assemblies

Outer wall Thermal transmittance $U_1 = 0.18$ W/(m² K)			Interstorey internal structure from steel beams		
No.	Material	T (cm)	No.	Material	T (cm)
1	Finishes (thin waterproofing layer)	–	1	Flooring–wooden parquet	1.5
2	Mineral wool in metal substructure	20.0	2	Concrete screed	5.0
3	Internal skeletal steel structure	22.0	3	Acoustic mineral wool	3.0
4	Secondary metal substructure	3.0	4	Steel substructure for final layers	3.0
5	Two layers of plasterboards	3.0	5	Primary steel beams	22.0
			6	Plasterboard	1.5

Appendix: Examples of the Use of the Methodology for Evaluating Structural Details

Building Physics

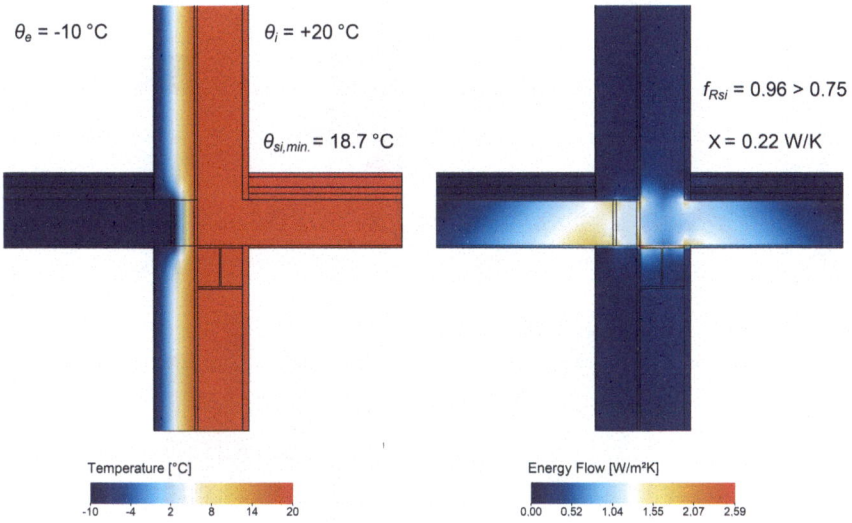

$\theta_e = -10\ °C$ $\theta_i = +20\ °C$

$\theta_{si,min.} = 18.7\ °C$

$f_{Rsi} = 0.96 > 0.75$

$X = 0.22\ W/K$

Temperature [°C]

Energy Flow [W/m²K]

Detail Evaluation

		Score	Selected weighting factors (weights)	Corrected weighting factors (influence of external factors)	Share of weighted score
Environmental & energy-efficiency parameters	E1	5	1	1	0.71
	E2	5	1	1	0.71
	E3	5	1	1	0.71
	E4	5	1	1	0.71
	E5	4	1	1	0.57
	E6	3	1	1	0.43
	E7	4	1	1	0.57
Technical & structural parameters	K1	4	1	1	0.57
	K2	2	1	1	0.29
	K3	4	1	1	0.57
	K4	3	1	1	0.43
	K5	5	1	1	0.71
	K6	3	1	1	0.43
	K7	4	1	1	0.57

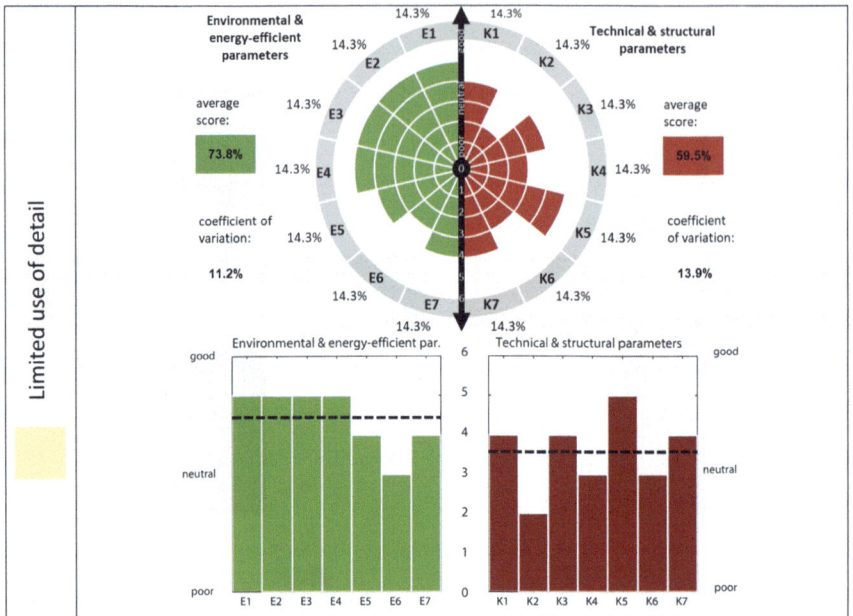

Concluding Remarks

A solution for a thermal bridge at the connection between a steel cantilever beam for the balcony and a steel beam inside the building was analysed. A precast element manages to eliminate the thermal bridge, supporting a good environmental and energy-efficiency assessment. In earthquake-prone areas, the detail must be symmetrical and have the load-bearing capacity for compression also in the upper half of the contact (a positive bending moment for seismic loads can occur). Precast load-bearing TI elements for steel beams comprise several modules. Modules that have the load-bearing capacity only for the tensile force also appear on the market. In earthquake-prone areas, it is recommended to use modules that also contain compression members (have the load-bearing capacity for compression and tension) and which must be placed in lower as well as in the upper part of the contact. This prevents damage to the contact with the cantilever uplift in the event of vertical seismic excitation. The detail can also be used in earthquake-prone areas, but only if the load-bearing insulation modules aiming to prevent thermal bridges at the contact of steel beams are used correctly.

Appendix: Examples of the Use of the Methodology for Evaluating Structural Details 167

Detail code	DB-03
Brief description	Building connection detail between a timber balcony structure and an RC building structure

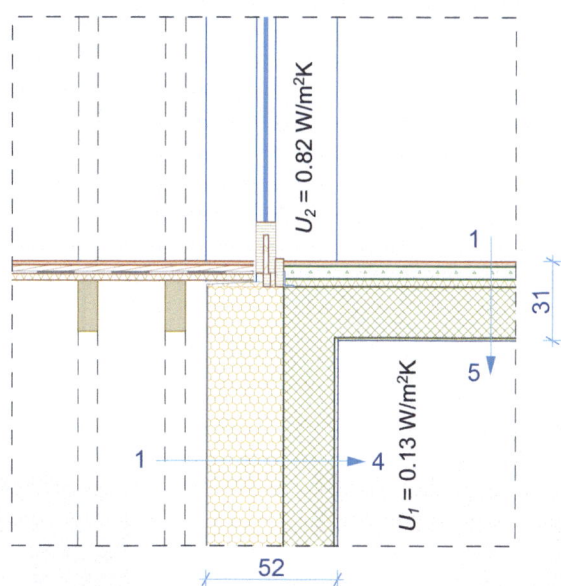

Detail Properties

Detail *DB*-03 shows a free-standing timber structure for the balcony anchored to the primary RC building structure. Also analysed is the contact between the balcony door and an RC outer wall and the interstorey slab. The highest heat losses can be expected through glazing and the balcony door frame, with additional losses occurring through steel anchors for the timber structure of the balcony (values for individual anchors):

- 0.011 W/K for steel ϕ 10 mm
- 0.004 W/K for stainless steel ϕ 10 mm.

The detail is composed of:

- RC outer wall insulated with EPS ($U_1 = 0.13$ W/(m² K))
- Balcony door ($U_2 = 0.82$ W/(m² K)).

Structural Assemblies

RC outer wall Thermal transmittance $U_1 = 0.13$ W/m² K			RC interstorey slab		
No.	Material	T (cm)	No.	Material	T (cm)
1	Silicate thin-layer plastering	–	1	Flooring–wooden parquet	1.5
2	EPS (expanded polystyrene)	30.0	2	Concrete screed	5.0
3	RC outer wall	20.0	3	Acoustic mineral wool	3.0
4	Internal plastering	1.5	4	RC load-bearing slab	20.0
			5	Internal plastering	1.5

Building Physics

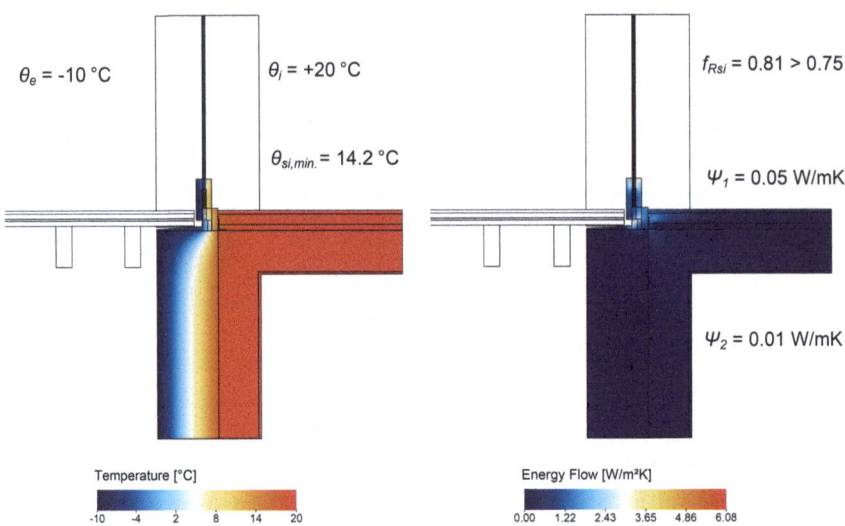

$\theta_e = -10\ °C$ $\theta_i = +20\ °C$ $\theta_{si,min.} = 14.2\ °C$ $f_{Rsi} = 0.81 > 0.75$ $\Psi_1 = 0.05$ W/mK $\Psi_2 = 0.01$ W/mK

Ψ_1...linear thermal transmittance in direction of the window frame
Ψ_2...linear thermal transmittance in direction of the façade

Detail Evaluation

		Score	Selected weighting factors (weights)	Corrected weighting factors (influence of external factors)	Share of weighted score
Environmental & energy-efficiency parameters	E1	6	1	1	0.86
	E2	6	1	1	0.86
	E3	5	1	1	0.71
	E4	5	1	1	0.71

(continued)

Appendix: Examples of the Use of the Methodology for Evaluating Structural Details   169

(continued)

	E5	4	1	1	0.57
	E6	4	1	1	0.57
	E7	3	1	1	0.43
Technical & structural parameters	K1	5	1	1	0.71
	K2	6	1	1	0.86
	K3	4	1	1	0.57
	K4	3	1	1	0.43
	K5	5	1	1	0.71
	K6	5	1	1	0.71
	K7	4	1	1	0.57

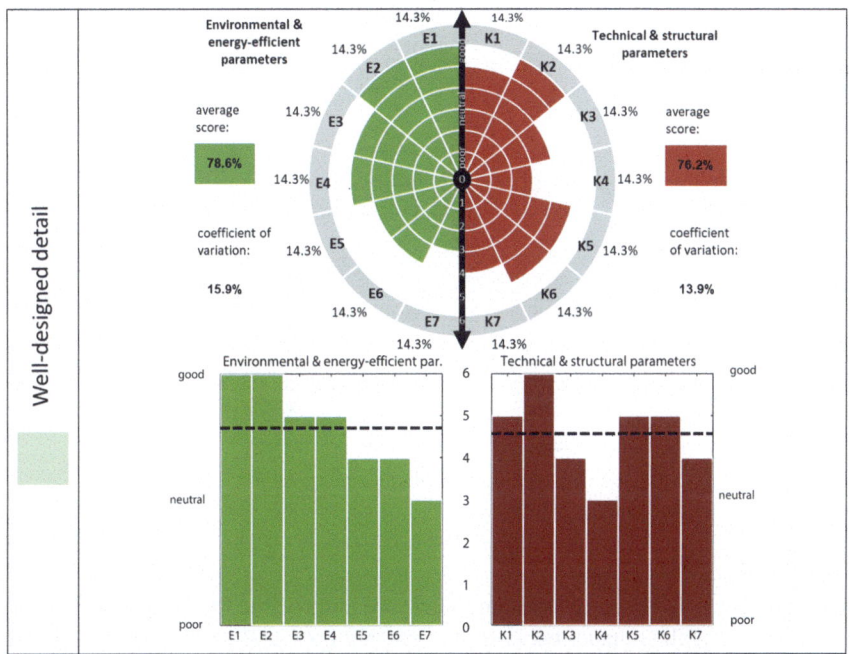

Concluding Remarks

The advantage of the analysed detail solution lies in the fact that there are no major heat losses due to the penetration of the RC cantilever slab. Thermal bridges are avoided by setting up a separate timber structure for the balcony next to the building. Such a solution is suitable for the construction of new buildings and renovations. In the latter, dilapidated RC cantilever slabs can be demolished and replaced with a new separate timber structure. In general, it is a good detail from the environmental and energy-efficiency as well as technical and structural aspects. Certain limitations regarding the durability of the timber structure (challenging realisation, exposure to external actions) must be taken into account, and attention must be paid to the anchorage of the timber structure on the primary RC structure. Insufficient anchorage could result in poorer resistance to horizontal seismic forces.

Detail code	*DB*-04
Brief description	Building connection detail between a hanging steel cantilever and an RC building structure

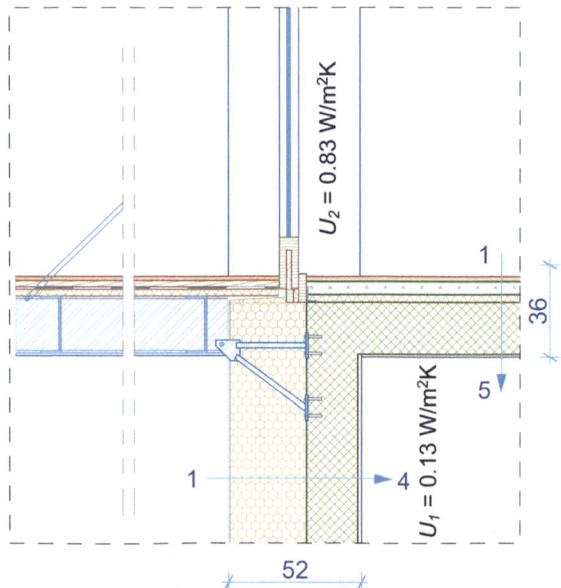

Detail Properties

Analysed here is the detail of a hanging steel cantilever on the RC building structure. Also analysed is the contact between the balcony door and an RC outer wall and the interstorey slab. The highest heat losses can be expected through glazing and the balcony door frame, with additional losses occurring through steel anchors for the hanging steel cantilever (values for individual anchors):

- 0.011 W/K for steel ϕ 10 mm
- 0.021 W/K for the three-legged steel angle.

The detail is composed of:

- RC outer wall insulated with EPS ($U_1 = 0.13$ W/(m² K))
- Balcony door ($U_2 = 0.82$ W/(m² K)).

Appendix: Examples of the Use of the Methodology for Evaluating Structural Details   171

Structural Assemblies

Outer wall Thermal transmittance $U_1 = 0.13$ W/(m² K)			RC interstorey slab		
No.	Material	T (cm)	No.	Material	T (cm)
1	Silicate thin-layer plastering	–	1	Flooring–wooden parquet	1.5
2	EPS (expanded polystyrene)	30.0	2	Concrete screed	5.0
3	RC outer wall	20.0	3	Acoustic mineral wool	3.0
4	Internal plastering	1.5	4	RC load-bearing slab	20.0
			5	Internal plastering	1.5

Building Physics

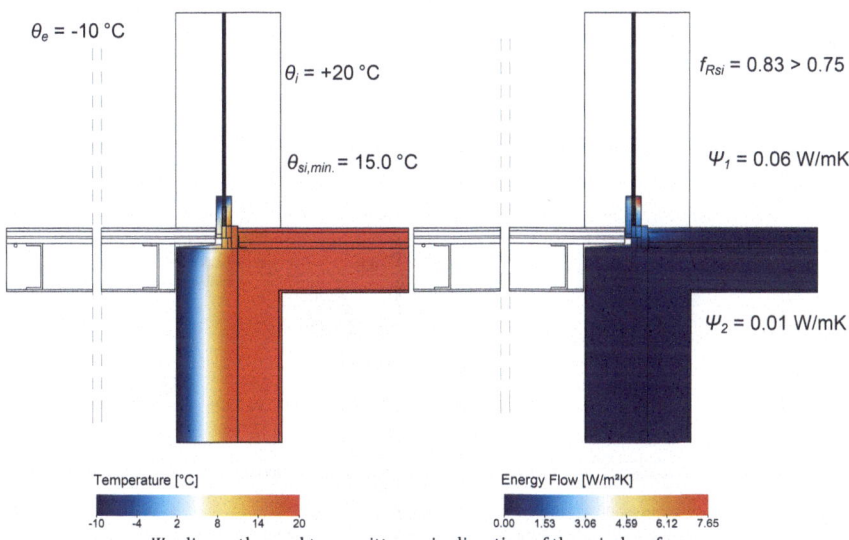

Ψ_1...linear thermal transmittance in direction of the window frame
Ψ_2...linear thermal transmittance in direction of the façade

Detail Evaluation

		Score	Selected weighting factors (weights)	Corrected weighting factors (influence of external factors)	Share of weighted score
Environmental & energy-efficiency parameters	E1	6	1	1	0.86
	E2	6	1	1	0.86
	E3	5	1	1	0.71
	E4	5	1	1	0.71
	E5	4	1	1	0.57
	E6	3	1	1	0.43
	E7	4	1	1	0.57
Technical & structural parameters	K1	4	1	1	0.57
	K2	6	1	1	0.86
	K3	1	1	1	0.14
	K4	3	1	1	0.43
	K5	4	1	1	0.57
	K6	5	1	1	0.71
	K7	4	1	1	0.57

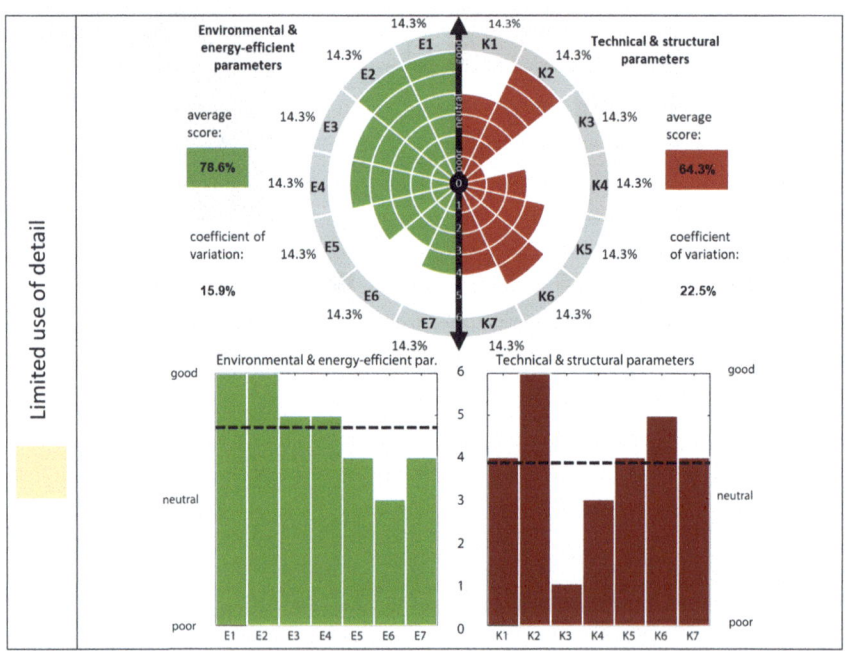

Concluding Remarks

Detail *DB*-04 shows a hanging cantilever with a metal substructure. The moment load-bearing capacity and sufficient anchorage must be ensured to make such a detail useful in earthquake-prone areas. Since the system is separated from the basic RC structure, its general technical and structural assessment is better than that of the RC cantilever with a precast load-bearing TI elements, as it does not significantly affect the load-bearing capacity of the primary RC building structure. Similarly to *DB*-03, the advantage of the detail is that there are no major heat losses due to the balcony structure, which result only from steel anchors for the hanging cantilever. Such a solution is suitable for the construction of new buildings and renovations. In the latter, existing uninsulated RC cantilever slabs can be demolished and replaced with a new separate hanging cantilever structure. In renovations, boundary conditions for the anchorage of the cantilever must be ensured and the influence of the cantilever as an additional load on the primary RC structure must be checked.

Detail code	*DK*-01
Brief description	Building connection detail between an RC outer wall and the unheated basement

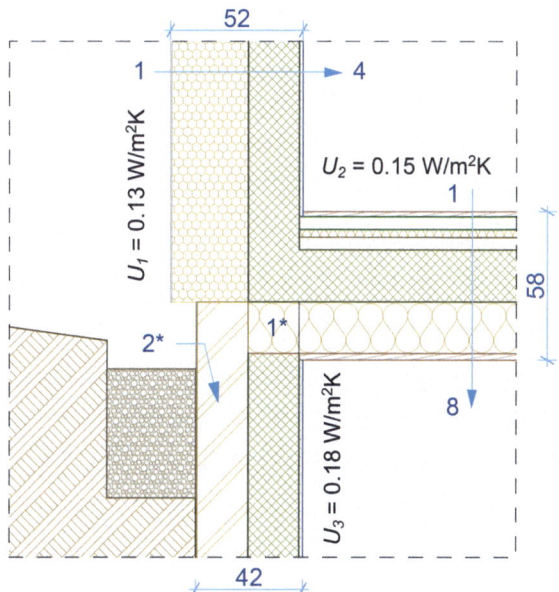

Detail Properties

The detail comprises the RC load-bearing structure. Its special feature is that the vertical load-bearing structure is interrupted at the location where the interstorey slab is in contact with the inserted TI (1*). TI does not interrupt the vertical RC structure along its whole length; instead, a certain predetermined percentage of penetrations is preserved (the load-bearing structure is preserved on 25% of the surface). The TI on the basement wall is at least 1 m below the underground section (2*).

The detail is composed of:

- RC outer wall insulated with EPS on the external side ($U_1 = 0.13$ W/(m² K))
- RC interstorey slab with min. wool on the bottom side ($U_2 = 0.15$ W/(m² K))
- RC basement wall insulated with XPS ($U_3 = 0.18$ W/(m² K)).

Structural Assemblies

Outer wall Thermal transmittance $U_1 = 0.13$ W/(m² K)			Interstorey slab between basement and ground floor Thermal transmittance $U_2 = 0.15$ W/(m² K)		
No.	Material	T (cm)	No.	Material	T (cm)
1	Silicate thin-layer plastering	–	1	Flooring–wooden parquet	1.5
2	EPS (expanded polystyrene)	32.0	2	Concrete screed	5.0
3	RC outer wall	20.0	3	PE vapour barrier	–
4	Finishes	–	4	Acoustic mineral wool	3.0
			5	Thermal insulation	5.0
			6	RC load-bearing slab	20.0
			7	Mineral wool in metal substructure	20.0
			8	Lightweight timber panels as a finish	2.5

Appendix: Examples of the Use of the Methodology for Evaluating Structural Details

Building Physics

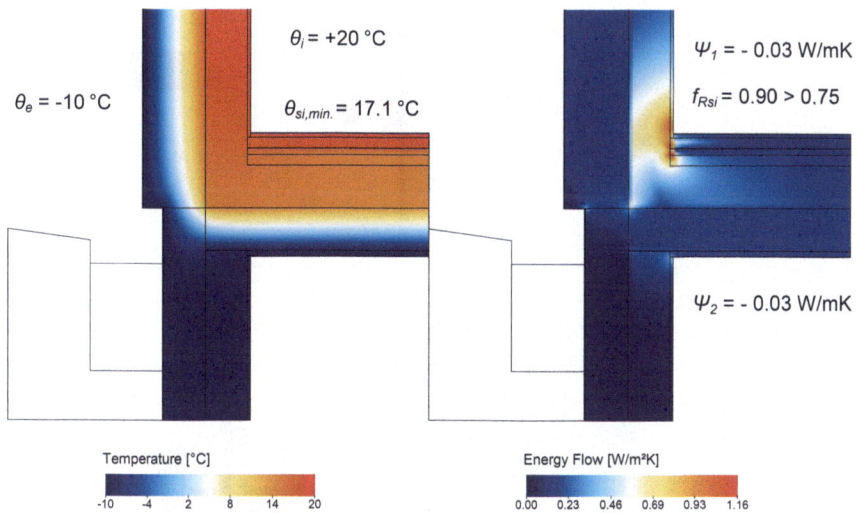

Ψ_1...linear thermal transmittance in direction of the external air
Ψ_2...linear thermal transmittance in direction of the unheated basement

Detail Evaluation

		Score	Selected weighting factors (weights)	Corrected weighting factors (influence of external factors)	Share of weighted score
Environmental & energy-efficiency parameters	E1	5	1	1	0.71
	E2	6	1	1	0.86
	E3	5	1	1	0.71
	E4	6	1	1	0.86
	E5	4	1	1	0.57
	E6	3	1	1	0.43
	E7	5	1	1	0.71
Technical & structural parameters	K1	0	1	1	0.00
	K2	1	1	1	0.14
	K3	1	1	1	0.14
	K4	1	1	1	0.14
	K5	3	1	1	0.43
	K6	1	1	1	0.14
	K7	2	1	1	0.29

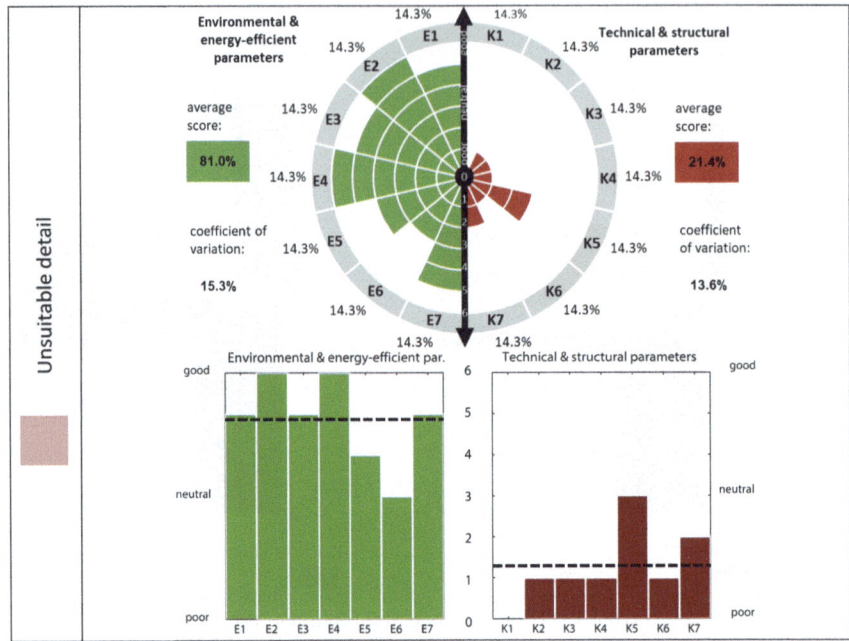

Concluding Remarks

The analysed solution of the contact between an outer wall and the unheated basement is not useful in earthquake-prone areas and must be replaced, since the penetration in thermal insulation is located in the critical region of the vertical RC structure. At this location, the load-bearing capacity and stiffness are reduced, there is no adhesion of steel rebars to concrete, etc. The load-bearing capacity of the detail largely depends on the length of the inserted thermal insulation (at what length the load-bearing structure is interrupted - in our case, the assumption is that only 25% of the load-bearing structure is continuous). At the same time, increased penetration of the load-bearing structure results in a more intensive heat flow and greater influence of the thermal bridge. From the energy aspect, the cross-section of the detail with the TI was analysed. At the location of the penetration of the vertical RC structure, the energy characteristics of the detail deteriorate. This can be corrected with external factors Z3 (influence on the global analysis) and Z5 (penetrations and openings).

Appendix: Examples of the Use of the Methodology for Evaluating Structural Details

Detail code	*DK*-02
Brief description	**Building connection detail between a cross-laminated timber (CLT) outer wall and the unheated basement**

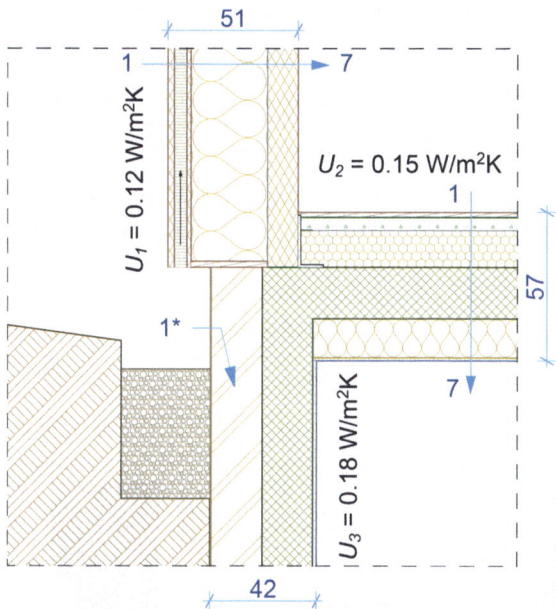

Detail Properties

The detail is composed of a CLT wall fixed to an RC interstorey slab by shear angle brackets and an RC basement wall. A thermal bridge resulting from the contact between the outer wall and the unheated basement is reduced on account of TI on the external and internal side of the RC slab. The timber structure is also shifted from the load-bearing axis to provide better continuity of TI. The TI on the basement wall is at least 1 m below the underground section (1*).

The detail is composed of:

- ventilated wooden façade fixed to a CLT wall and insulated with mineral wool ($U_1 = 0.12$ W/(m² K))
- RC interstorey slab with TI in the cover ($U_2 = 0.15$ W/(m² K))
- basement wall from concrete hollow bricks insulated with XPS ($U_3 = 0.18$ W/(m² K)).

Structural Assemblies

Outer wall Thermal transmittance $U_1 = 0.12$ W/(m² K)			Interstorey slab between basement and ground floor Thermal transmittance $U_2 = 0.15$ W/(m² K)		
No.	Material	T (cm)	No.	Material	T (cm)
1	Wood panelling	2.5	1	Flooring–wooden parquet	2.0
2	Ventilated layer	5.0	2	Concrete screed	5.0
3	MDF	1.6	3	PE vapour barrier	0.1
4	Mineral wool in a timber substructure	30.0	4	XPS board	14.0
5	PE vapour barrier	–	5	RC load-bearing slab	20.0
6	Wall from cross-laminated timber panels	12.0	6	Mineral wool in metal substructure	15.0
7	Finishes	–	7	Plasterboard	1.5

Building Physics

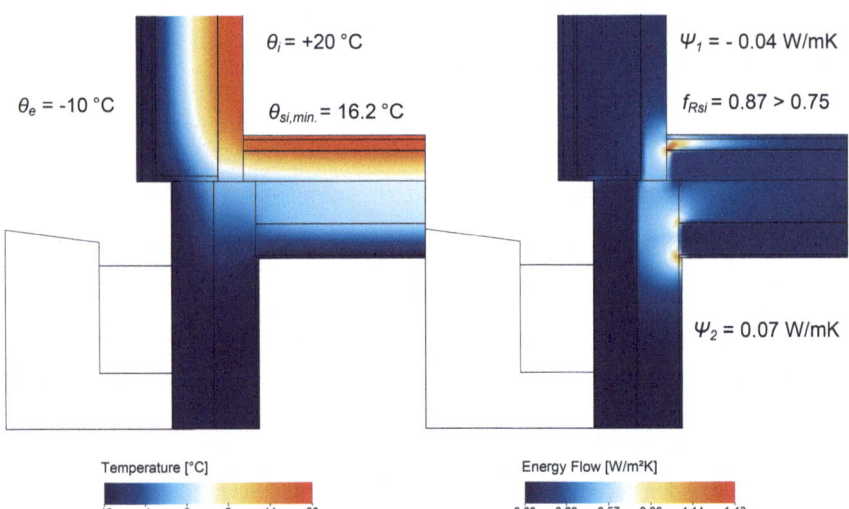

$\theta_i = +20\ °C$
$\theta_e = -10\ °C$
$\theta_{si,min.} = 16.2\ °C$
$\Psi_1 = -0.04$ W/mK
$f_{Rsi} = 0.87 > 0.75$
$\Psi_2 = 0.07$ W/mK

Temperature [°C]
Energy Flow [W/m²K]

Ψ_1...linear thermal transmittance in direction of the external air
Ψ_2...linear thermal transmittance in direction of the unheated basement

Appendix: Examples of the Use of the Methodology for Evaluating Structural Details 179

Detail Evaluation

		Score	Selected weighting factors (weights)	Corrected weighting factors (influence of external factors)	Share of weighted score
Environmental & energy-efficiency parameters	E1	5	1	1	0.71
	E2	3	1	1	0.43
	E3	4	1	1	0.57
	E4	5	1	1	0.71
	E5	3	1	1	0.43
	E6	4	1	1	0.57
	E7	2	1	1	0.29
Technical & structural parameters	K1	3	1	1	0.43
	K2	3	1	1	0.43
	K3	2	1	1	0.29
	K4	3	1	1	0.43
	K5	2	1	1	0.29
	K6	3	1	1	0.43
	K7	3	1	1	0.43

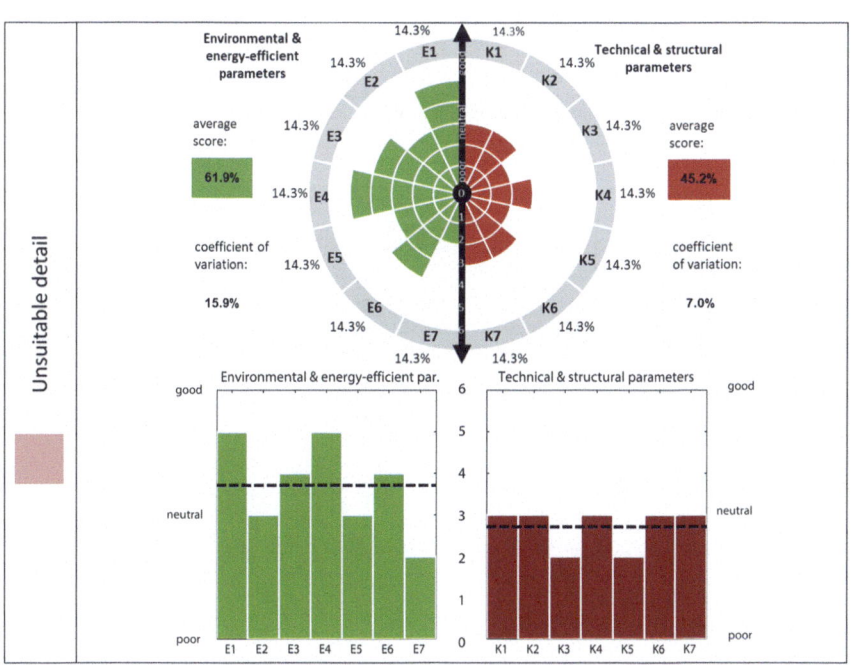

Concluding Remarks

The analysed detail shows the contact between an outer wall and the unheated basement. The solution for a thermal bridge is thermal insulation on the bottom side of the basement slab and a shift in the vertical load-bearing structure. Such a form of the detail could be improved for use in earthquake-prone areas, as it should be symmetrical without any shifts in the vertical timber load-bearing structure and also the CLT wall should be symmetrically fixed to the RC foundation with a combination of shear angle brackets and hold-downs. This solution could result in a poorer continuity of TI and greater heat flow towards the unheated basement and the outside. From the energy aspect, the detail is average, not at risk of mould, but also does not provide complete thermal comfort, since the inner surface temperature is 4 °C lower than the indoor air temperature.

Detail code	*DK*-03
Brief description	**Building connection detail between an outer light-frame timber wall and the unheated basement**

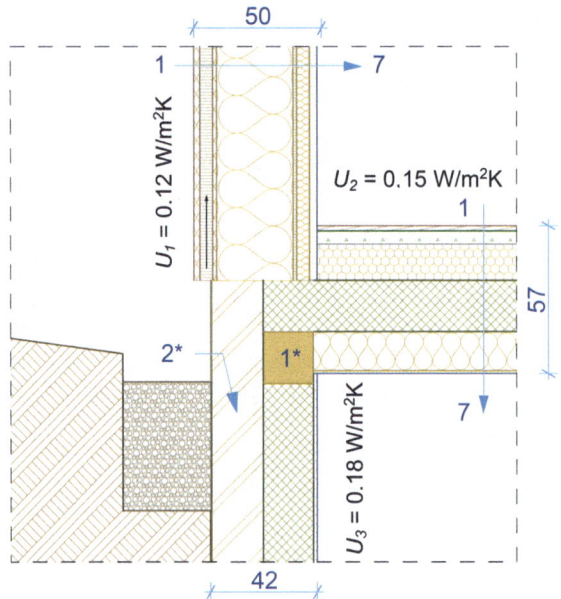

Detail Properties

The detail is composed of a light-frame timber wall, an RC basement wall, and an RC interstorey slab. A thermal bridge resulting from the contact between the outer wall and the unheated basement is prevented by an insulation base block located just below the location where the RC slab is fixed to the RC basement wall (1*). The TI on the basement wall is at least 1 m below the ground (2*).

The detail is composed of:

- ventilated wooden façade on a light-frame timber panel insulated with mineral wool ($U_1 = 0.12$ W/(m² K))
- RC interstorey slab with TI in the cover ($U_2 = 0.15$ W/(m² K))
- basement wall from concrete hollow bricks insulated with XPS ($U_3 = 0.18$ W/(m² K)).

Structural Assemblies

Outer wall Thermal transmittance $U_1 = 0.12$ W/(m² K)			Interstorey slab between basement and ground floor Thermal transmittance $U_2 = 0.15$ W/(m² K)		
No.	Material	T (cm)	No.	Material	T (cm)
1	Wood panelling	2.5	1	Flooring–wooden parquet	2.0
2	Ventilated layer	5.0	2	Concrete screed	5.0
3	MDF	1.6	3	PE vapour barrier	0.1
4	Mineral wool in timber substructure	30.0	4	XPS board	14.0
5	OSB board	1.8	5	RC load-bearing slab	20.0
6	Mineral wool in timber substructure	5.0	6	Mineral wool in metal substructure	15.0
7	Two layers of plasterboards	3.0	7	Plasterboard	1.5

Appendix: Examples of the Use of the Methodology for Evaluating Structural Details

Building Physics

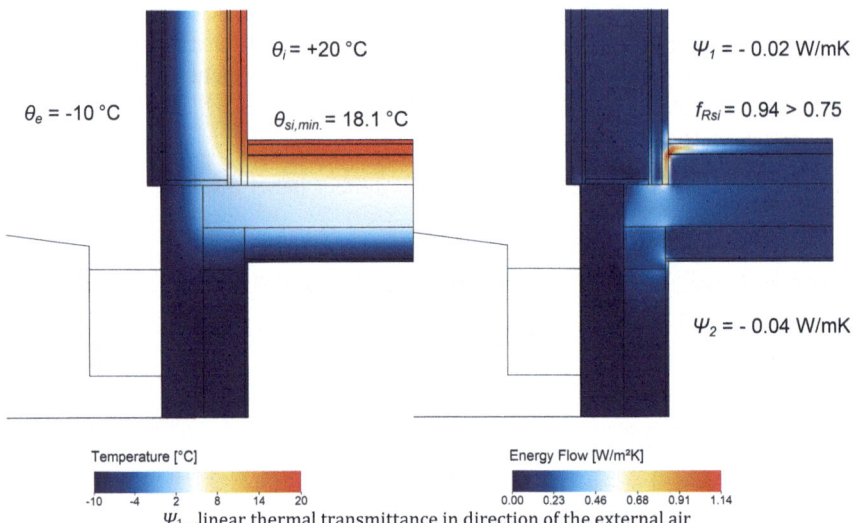

Ψ_1...linear thermal transmittance in direction of the external air
Ψ_2...linear thermal transmittance in direction of the unheated basement

Detail Evaluation

		Score	Selected weighting factors (weights)	Corrected weighting factors (influence of external factors)	Share of weighted score
Environmental & energy-efficiency parameters	E1	5	1	1	0.71
	E2	5	1	1	0.71
	E3	5	1	1	0.71
	E4	6	1	1	0.86
	E5	3	1	1	0.43
	E6	4	1	1	0.57
	E7	2	1	1	0.29
Technical & structural parameters	K1	1	1	1	0.14
	K2	1	1	1	0.14
	K3	1	1	1	0.14
	K4	0	1	1	0.00
	K5	0	1	1	0.00
	K6	2	1	1	0.29
	K7	2	1	1	0.29

Appendix: Examples of the Use of the Methodology for Evaluating Structural Details   183

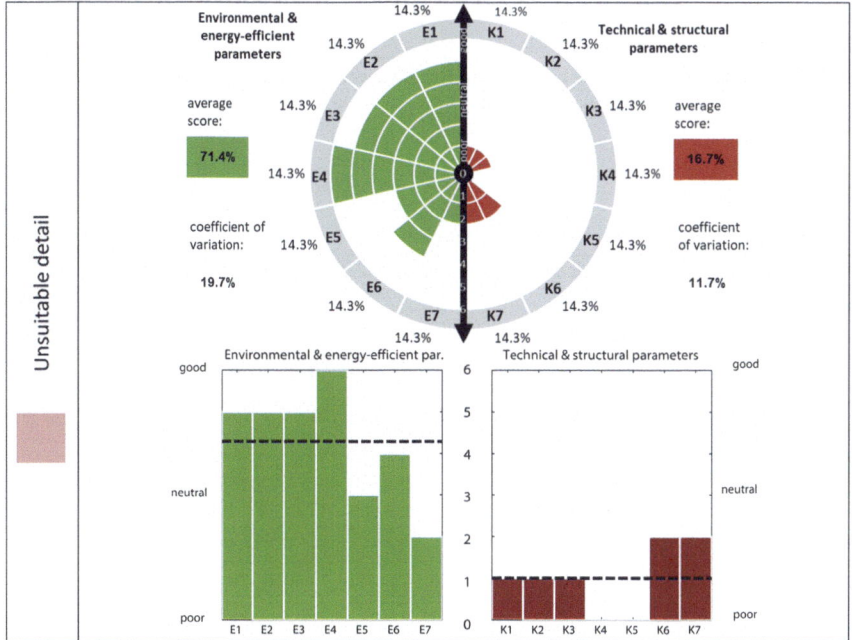

Concluding Remarks

The analysed detail shows the contact between a light-frame timber wall and the unheated basement. The solution for a thermal bridge is the use of TI on the bottom side of the basement slab and the installation of an insulation base block. Such a detail can be used only conditionally in earthquake-prone areas, as it must be symmetrical without any shifts in the vertical timber load-bearing structure, which would result in a greater thermal bridge, which is why such a contact is not possible from the energy aspect. Also recommended is the construction without the insulation base block which prevents the continuity of the vertical RC structure.

Detail code	*DK-04*
Brief description	**Building connection detail between an outer masonry wall and the unheated basement**

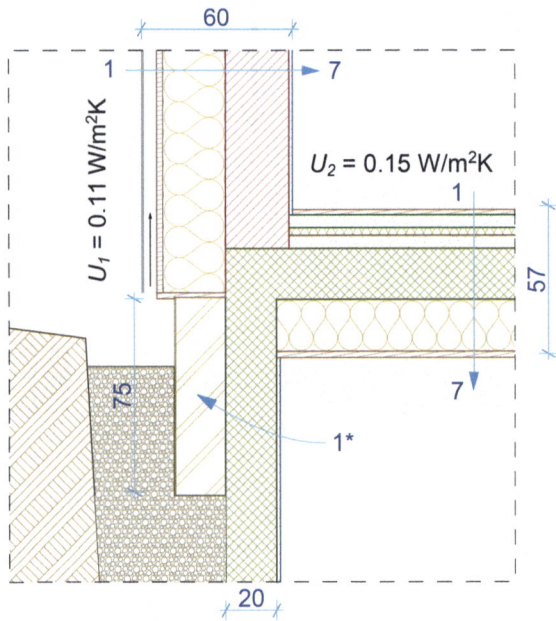

Detail Properties

The detail is composed of an outer masonry wall, an RC basement wall, and an RC interstorey slab. A thermal bridge resulting from the contact between the outer wall and the unheated basement is reduced on account of vertical TI 75 cm from the critical detail (1*).

The detail is composed of:

- outer masonry wall insulated with mineral wool ($U_1 = 0.11$ W/(m² K))
- RC interstorey slab with min. wool on the bottom side ($U_2 = 0.15$ W/(m² K))
- basement wall from concrete hollow bricks (uninsulated).

Appendix: Examples of the Use of the Methodology for Evaluating Structural Details

Structural Assemblies

Outer wall Thermal transmittance $U_1 = 0.11$ W/(m² K)			Interstorey slab between basement and ground floor Thermal transmittance $U_2 = 0.16$ W/(m² K)		
No.	Material	T (cm)	No.	Material	T (cm)
1	Fibre cement boards	2.5	1	Flooring–wooden parquet	1.5
2	Ventilated layer	5.0	2	Concrete screed	5.0
3	Wind barrier (felt, geotextile)	1.6	3	PE vapour barrier	–
4	Plywood	30.0	4	Acoustic mineral wool	3.0
5	Stone wool in timber substructure	1.8	5	Thermal insulation	5.0
6	Hollow brick (Poroblock 29/25)	5.0	6	RC load-bearing slab	20.0
7	Cement mortar	3.0	7	Mineral wool in metal substructure	20.0

Building Physics

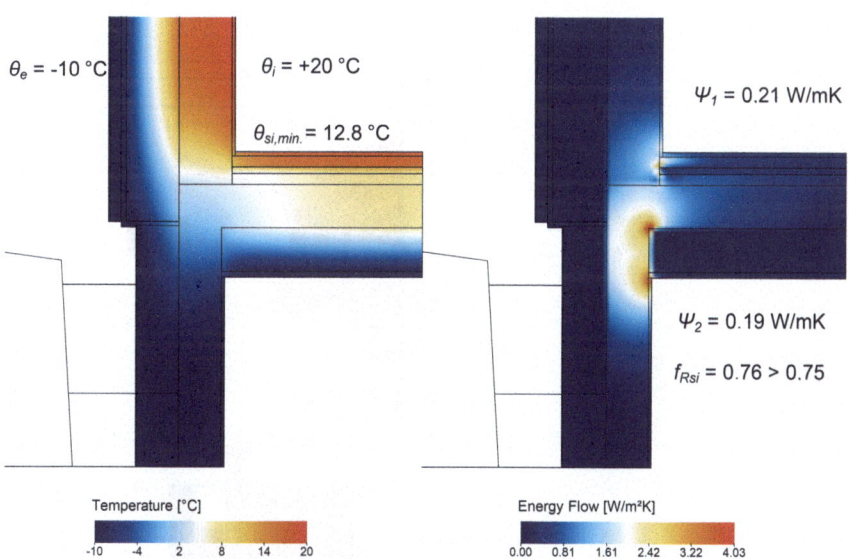

Ψ_1...linear thermal transmittance in direction of the external air
Ψ_2...linear thermal transmittance in direction of the unheated basement

Appendix: Examples of the Use of the Methodology for Evaluating Structural Details

Detail Evaluation

		Score	Selected weighting factors (weights)	Corrected weighting factors (influence of external factors)	Share of weighted score
Environmental & energy-efficiency parameters	E1	4	1	1	0.57
	E2	0	1	1	0.00
	E3	4	1	1	0.57
	E4	4	1	1	0.57
	E5	5	1	1	0.71
	E6	3	1	1	0.43
	E7	5	1	1	0.71
Technical & structural parameters	K1	5	1	1	0.71
	K2	5	1	1	0.71
	K3	5	1	1	0.71
	K4	5	1	1	0.71
	K5	6	1	1	0.86
	K6	4	1	1	0.57
	K7	5	1	1	0.71

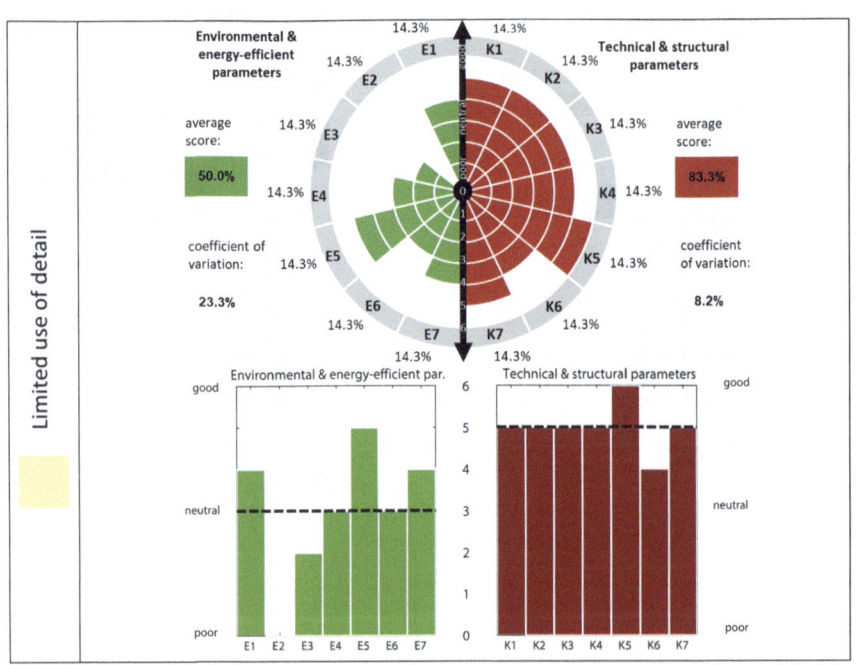

Concluding Remarks

From the structural aspect, the analysed solution is suitable, as there are no interruptions of the load-bearing structure, apart from material replacement. Thermal bridges are prevented by extended vertical thermal insulation below the ground. This makes the detail only conditionally useful from the environmental and energy-efficiency aspect, since it allows a thermal bridge towards the unheated basement and the soil. Extending thermal insulation towards the basement wall could reduce the thermal bridge, but a thermal bridge towards the unheated basement (thermal insulation is not continuous at this location) could still not be avoided. Heat losses resulting from the thermal bridge of the analysed detail must be taken into account in the calculation of the use of energy in the building.

Detail code	*DK*-05
Brief description	Building connection detail between an RC outer wall and a heated basement

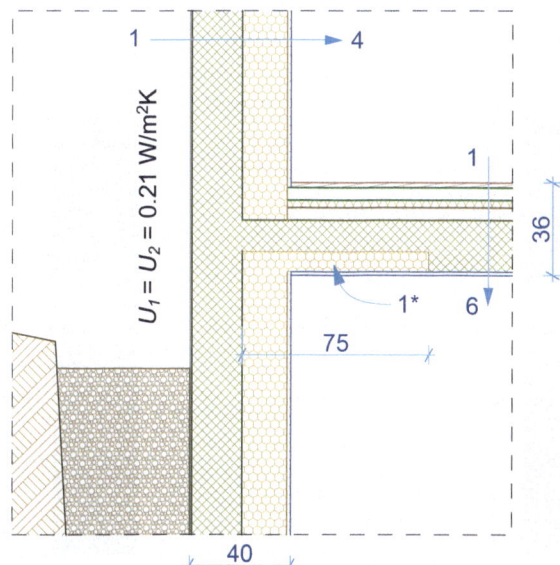

Detail Properties

The detail is composed of an RC outer and an RC basement wall (TI on the internal side), and an RC interstorey slab. A thermal bridge resulting from the installation of TI on the internal side is reduced by installing TI at the location where the RC interstorey slab is fixed (1*). The length of TI from the location where the RC interstorey slab is fixed is 75 cm and its thickness is 8 cm.

The detail is composed of:

- RC outer and basement wall insulated with EPS ($U_1 = U_2 = 0.21$ W/(m² K))
- RC interstorey slab insulated with EPS at the location where it is fixed.

Structural Assemblies

Outer wall Thermal transmittance $U_1 = 0.21$ W/(m² K)			Interstorey slab between basement and ground floor		
No.	Material	T (cm)	No.	Material	T (cm)
1	Silicate thin-layer plastering	–	1	Flooring–wooden parquet	1.5
2	RC outer wall	20.0	2	Concrete screed	5.0
3	EPS (expanded polystyrene)	20.0	3	Acoustic mineral wool	3.0
4	Finishes	–	4	Thermal insulation	5.0
			5	RC load-bearing slab	20.0
			6	Finishes	–

Building Physics

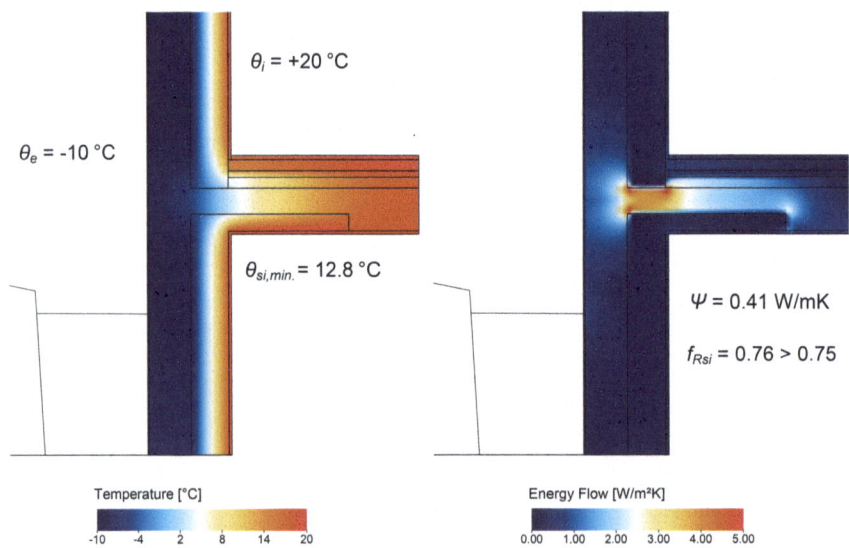

$\theta_e = -10\ °C$
$\theta_i = +20\ °C$
$\theta_{si,min.} = 12.8\ °C$
$\Psi = 0.41$ W/mK
$f_{Rsi} = 0.76 > 0.75$

Temperature [°C]
Energy Flow [W/m²K]

Appendix: Examples of the Use of the Methodology for Evaluating Structural Details   189

Detail Evaluation

		Score	Selected weighting factors (weights)	Corrected weighting factors (influence of external factors)	Share of weighted score
Environmental & energy-efficiency parameters	E1	4	1	1	0.57
	E2	2	1	1	0.29
	E3	2	1	1	0.29
	E4	1	1	1	0.14
	E5	5	1	1	0.71
	E6	2	1	1	0.29
	E7	5	1	1	0.71
Technical & structural parameters	K1	1	1	1	0.14
	K2	2	1	1	0.29
	K3	3	1	1	0.43
	K4	3	1	1	0.43
	K5	4	1	1	0.57
	K6	3	1	1	0.43
	K7	4	1	1	0.57

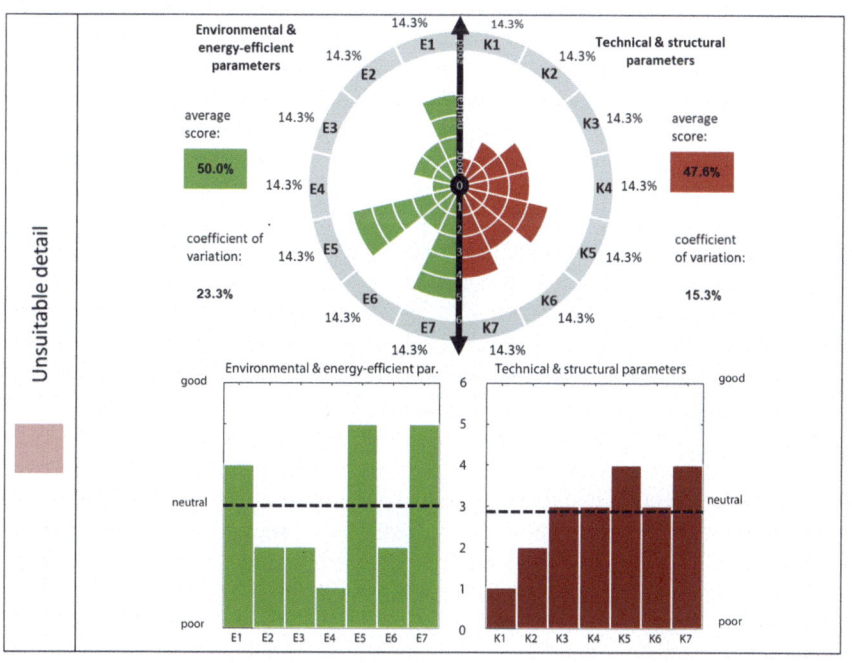

Concluding Remarks

The proposed solution for the thermal bridge resulting from the installation of TI on the internal side of the outer wall is only conditionally useful from the technical and structural, and environmental and energy-efficiency aspects. From the structural aspect, the greatest weakness is strength deterioration due to the installation of TI, which is located in the critical region next to where the RC slab is fixed on the outer wall. At this location, the possibility for denser installation of steel reinforcement, as well as its symmetrical or continued alignment is reduced, also reducing the load-bearing capacity and stiffness of the detail. The installation of TI does not suitably solve the thermal bridge, since the temperature range is still unfavourable. The temperature on the inner surface border the critical temperature, which could lead to condensation and mould.

Detail code	*DS*-01
Brief description	Building connection detail between a timber roof and an outer wall

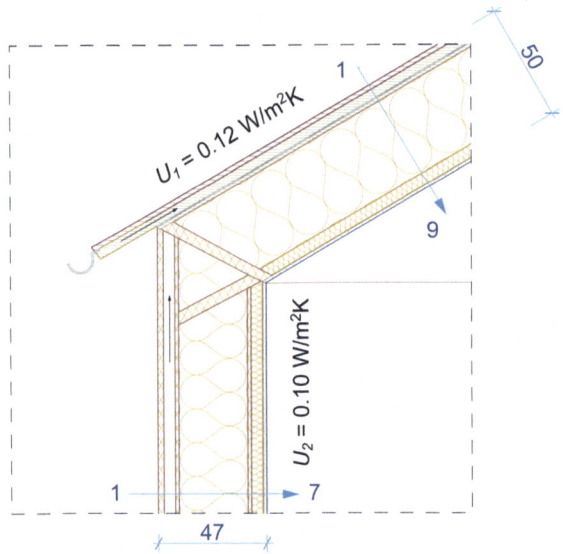

Detail Properties

The building connection detail between the roof and an outer wall for light-frame timber structure is analysed. The use of a frame-type structure enables a reduction of the geometrical thermal bridge, since TI is almost completely uninterrupted. TI is only interrupted at locations of timber beams in the roof and other load-bearing elements.

Appendix: Examples of the Use of the Methodology for Evaluating Structural Details

The detail is composed of:

- pitched timber roof structure insulated with cellulose TI ($U_1 = 0.12$ W/(m² K))
- outer light-frame timber wall insulated with wood wool ($U_2 = 0.10$ W/(m² K)).

Structural Assemblies

Pitched roof Thermal transmittance $U_1 = 0.12$ W/(m² K)				Outer wall Thermal transmittance $U_2 = 0.10$ W/(m² K)			
No.	Material	T (cm)		No.	Material	T (cm)	
1	Sheet metal roofing	0.5		1	Wood panelling	2.5	
2	Timber substructure	2.0		2	Ventilated layer	5.0	
3	Ventilated layer	4.0		3	MDF	1.6	
4	Secondary roofing	–		4	Wood wool in timber substructure	30.0	
5	MDF	1.6		5	OSB board	1.8	
6	Cellulose TI in timber substructure	36.0		6	Sheep wool in timber substructure	5.0	
7	PE vapour barrier	–		7	Plasterboard	1.5	
8	Sheep wool in timber substructure	5.0					
9	Plasterboard	1.5					

Building Physics

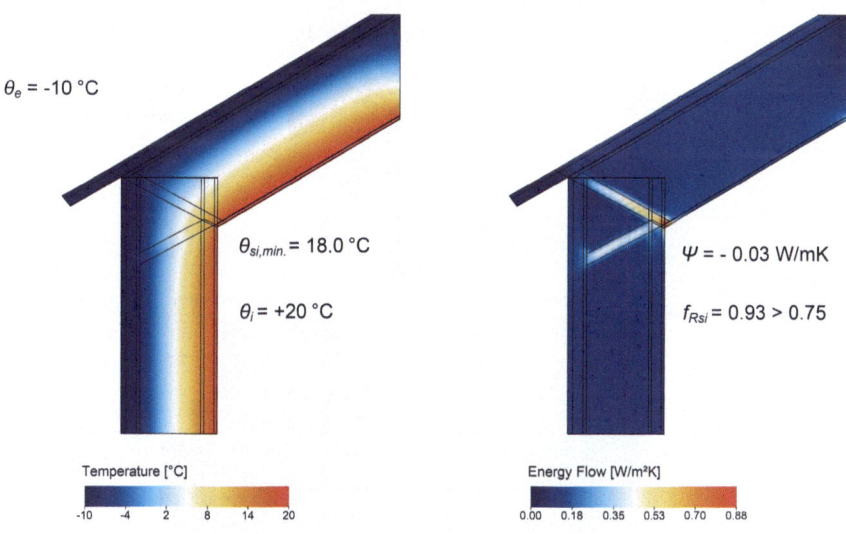

$\theta_e = -10\ °C$

$\theta_{si,min.} = 18.0\ °C$

$\theta_i = +20\ °C$

$\Psi = -0.03$ W/mK

$f_{Rsi} = 0.93 > 0.75$

Temperature [°C]

Energy Flow [W/m²K]

Appendix: Examples of the Use of the Methodology for Evaluating Structural Details

Detail Evaluation

		Score	Selected weighting factors (weights)	Corrected weighting factors (influence of external factors)	Share of weighted score
Environmental & energy-efficiency parameters	E1	6	1	1	0.86
	E2	5	1	1	0.71
	E3	6	1	1	0.86
	E4	6	1	1	0.86
	E5	3	1	1	0.43
	E6	6	1	1	0.86
	E7	2	1	1	0.29
Technical & structural parameters	K1	5	1	1	0.71
	K2	6	1	1	0.86
	K3	6	1	1	0.86
	K4	5	1	1	0.71
	K5	6	1	1	0.86
	K6	4	1	1	0.57
	K7	3	1	1	0.43

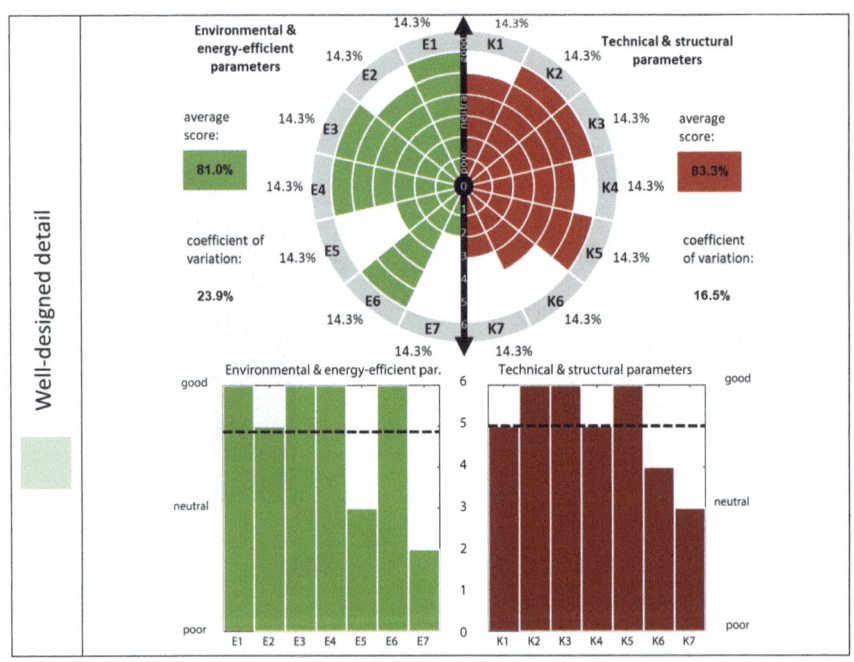

Concluding Remarks

According to most criteria, the analysed detail of the light-frame timber load-bearing structure received good scores, and is therefore recommended to be used while taking into account limitations of the construction of timber structures in earthquake-prone areas. The ductility of connectors and the symmetrical design should be ensured, and the limitations of timber regarding its strength and ductility (the detail cannot be used in all types of structures - it is limited by, for example, the number of storeys, etc.) must be taken into account. From the environmental and energy-efficiency aspect, the detail's score regarding airtightness and durability was lower, since it requires several protective layers to achieve optimal functioning and durability. The quality of the detail largely depends on the precision of construction (considering the contractor's experience, it can be taken into account with external parameter Z4).

Detail code	*DS*-02
Brief description	**Building connection detail between a cross-laminated timber (CLT) roof and CLT outer wall**

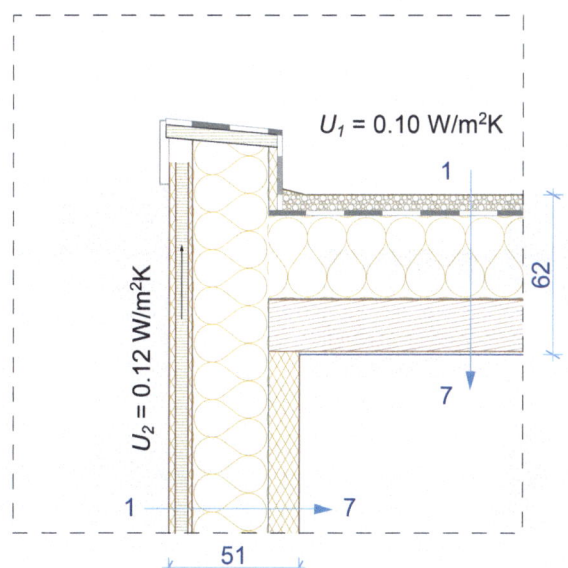

Detail Properties

The building connection detail between the CLT roof and CLT outer wall is analysed. TI is installed on the external side and is continuous. Therefore, no thermal bridges are expected. A special timber substructure is required to fix TI on the parapet.

The detail is composed of:

- CLT roof structure insulated with mineral wool ($U_1 = 0.10$ W/(m² K))
- outer CLT wall insulated with mineral wool ($U_2 = 0.12$ W/(m² K)).

Structural Assemblies

Flat roof Thermal transmittance $U_1 = 0.10$ W/(m² K)			Outer wall Thermal transmittance $U_2 = 0.12$ W/(m² K)		
No.	Material	T (cm)	No.	Material	T (cm)
1	Gravel 16/32	6.0	1	Wood panelling	2.5
2	Bitumen cardboard as waterproofing in two layers	1.0	2	Ventilated layer	5.0
3	Mineral wool board	32.0	3	MDF	1.6
4	ALU-bitumen layer	–	4	Min. wool in timber substructure	30.0
5	Vapour barrier	–	5	PE vapour barrier	–
6	CLT panels	20.0	6	CLT wall panels	12.0
7	Two layers of plasterboards	3.0	7	Finishes	–

Building Physics

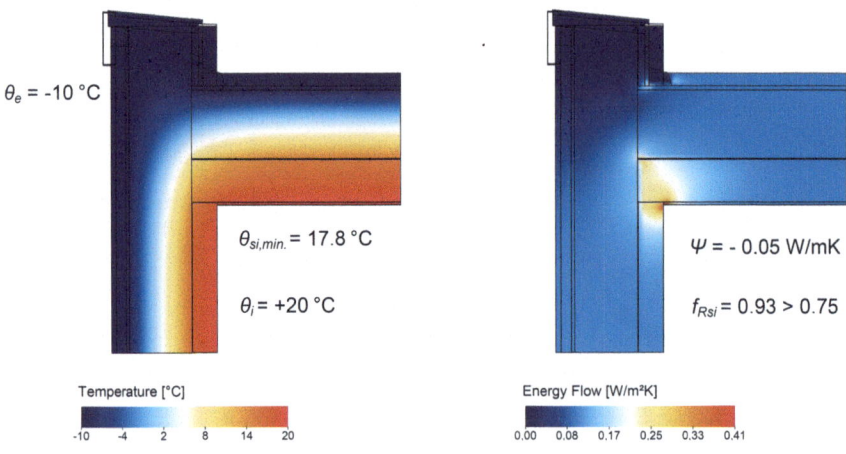

$\theta_e = -10$ °C

$\theta_{si,min.} = 17.8$ °C

$\theta_i = +20$ °C

$\Psi = -0.05$ W/mK

$f_{Rsi} = 0.93 > 0.75$

Temperature [°C]

Energy Flow [W/m²K]

Appendix: Examples of the Use of the Methodology for Evaluating Structural Details

Detail Evaluation

		Score	Selected weighting factors (weights)	Corrected weighting factors (influence of external factors)	Share of weighted score
Environmental & energy-efficiency parameters	E1	6	1	1	0.86
	E2	5	1	1	0.71
	E3	6	1	1	0.86
	E4	6	1	1	0.86
	E5	3	1	1	0.43
	E6	6	1	1	0.86
	E7	2	1	1	0.29
Technical & structural parameters	K1	5	1	1	0.71
	K2	5	1	1	0.71
	K3	4	1	1	0.57
	K4	5	1	1	0.71
	K5	5	1	1	0.71
	K6	3	1	1	0.43
	K7	3	1	1	0.43

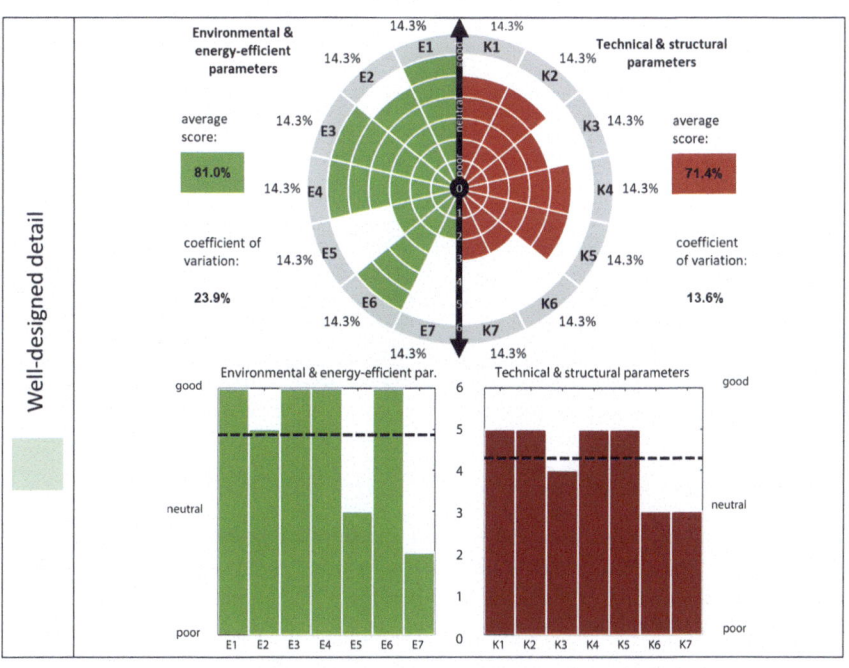

Concluding Remarks

The building connection between two structural assemblies with a CLT load-bearing structure is typically a good detail also in earthquake-prone areas. However, certain limitations must be considered (e.g. whether the connectors are appropriately selected (combination of hold-downs and angle brackets), symmetrical and ductile, limitations of the strength and ductility of timber, limited number of storeys, etc.). The main role among the environmental and energy-efficiency parameters is assumed by the protection of thermal insulation and the mass timber load-bearing structure against external actions. The detail's score is lower due to the complexity and precision of the detail construction, which are crucial to optimal and durable functioning of the detail. The quality of the detail largely depends on the precision of construction (based on the contractor's experience, it can be taken into account with external parameter Z4).

Detail code	DS-03
Brief description	Building connection detail between a timber roof and an outer masonry wall

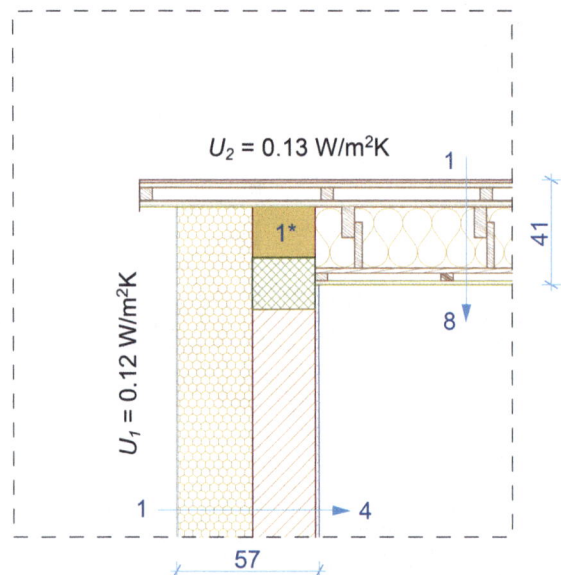

Detail Properties

The building connection detail between a single-pitch timber roof and an outer masonry wall considered in the longitudinal direction is analysed. A thermal bridge occurs, which is reduced by inserting load-bearing TI (1*) with thermal conductivity $\lambda = 0.04$ W/(m K). Timber roof beams are composed of two parts, which are offset in order to prevent local thermal bridges in the timber roof structural assembly.

The detail is composed of:

- outer masonry wall insulated with EPS ($U_1 = 0.12$ W/(m² K))
- timber roof structure insulated with mineral wool ($U_2 = 0.13$ W/(m² K)).

Structural Assemblies

Outer wall			Single-pitch roof		
Thermal transmittance $U_1 = 0.12$ W/(m² K)			Thermal transmittance $U_2 = 0.13$ W/(m² K)		
No.	Material	T (cm)	No.	Material	T (cm)
1	Silicate thin-layer plastering	–	1	Sheet metal roofing	0.5
2	EPS (expanded polystyrene)	30.0	2	Timber substructure	2.4
3	Outer masonry wall	25.0	3	Ventilated layer	5.0
4	Cement mortar	1.5	4	Secondary roofing/wind barrier	–
			5	Timber substructure	2.4
			6	Min. wool in timber substructure	24.0
			7	Timber substructure and air layer	5.0
			8	Plasterboard	1.5

Building Physics

$\theta_e = -10$ °C

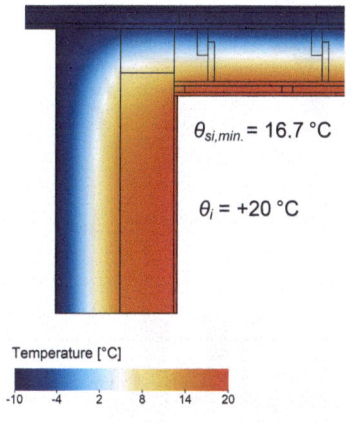

$\theta_{si,min.} = 16.7$ °C

$\theta_i = +20$ °C

Temperature [°C]

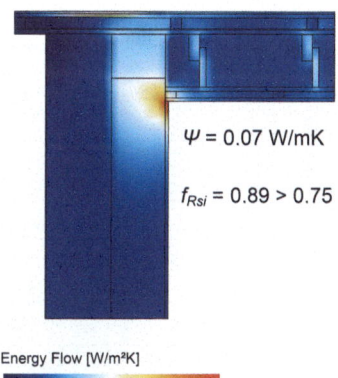

$\Psi = 0.07$ W/mK

$f_{Rsi} = 0.89 > 0.75$

Energy Flow [W/m²K]

Detail Evaluation

		Score	Selected weighting factors (weights)	Corrected weighting factors (influence of external factors)	Share of weighted score
Environmental & energy-efficiency parameters	E1	6	1	1	0.86
	E2	5	1	1	0.71
	E3	5	1	1	0.71
	E4	5	1	1	0.71
	E5	3	1	1	0.43
	E6	4	1	1	0.57
	E7	3	1	1	0.43
Technical & structural parameters	K1	3	1	1	0.43
	K2	3	1	1	0.43
	K3	4	1	1	0.57
	K4	4	1	1	0.57
	K5	3	1	1	0.43
	K6	4	1	1	0.57
	K7	3	1	1	0.43

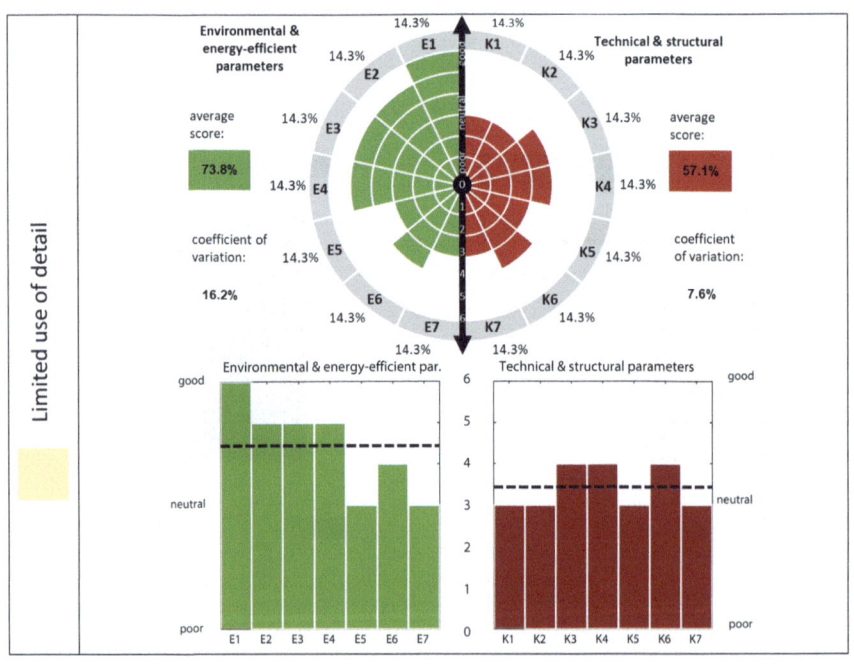

Appendix: Examples of the Use of the Methodology for Evaluating Structural Details   199

Concluding Remarks

The analysed building connection detail is located at the contact between the single-pitch roof structure and an outer wall (a cross section in direction of the slope of the roof is analysed). The detail also shows that the roof beam is composed of two parts, resulting in lower bending strength and more difficult fixing on the outer wall. The beams are composed of two parts and shifted to reduce the heat flow which is located higher on the timber beams. In addition to this measure, a thermal bridge through the outer wall is prevented with load-bearing thermal insulation (e.g. XPS). The anchorage of all roof substructures is made difficult at this location. The total score of such a detail is lower from the technical and structural aspect. From the environmental and energy-efficiency aspect, the detail is not problematic and scored well with a slight deduction from the aspect of sustainability and airtightness.

Detail code	DS-04
Brief description	Building connection detail between an RC roof and an outer masonry wall with a parapet

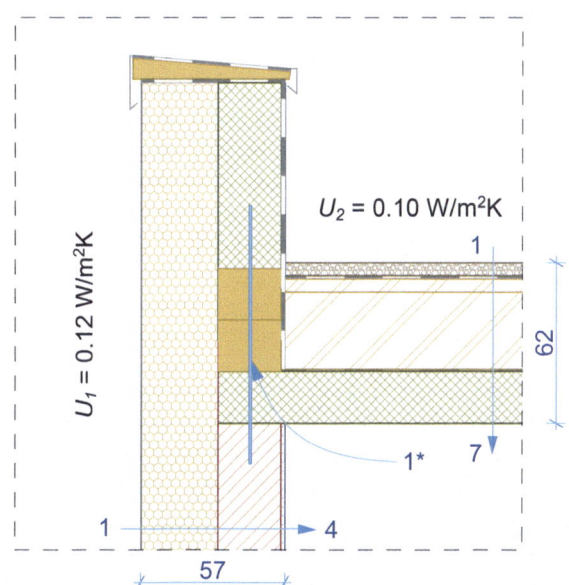

Detail Properties

The building connection detail between an RC flat roof and an outer masonry wall is analysed. The outer wall is extended with a parapet above the benchmark of the axis of the load-bearing RC roof slab. The thermal insulation element (1*) with thermal conductivity $\lambda = 0.12$ W/(m K) (e.g. autoclaved aerated concrete) prevents a thermal bridge resulting from heat flow through the RC structure of the parapet. The parapet is anchored along the axis of the vertical load-bearing structure.

The detail is composed of:

- outer masonry wall insulated with EPS ($U_1 = 0.12$ W/(m² K))
- RC roof slab insulated with XPS ($U_2 = 0.10$ W/(m² K)).

Structural Assemblies

Outer wall Thermal transmittance $U_1 = 0.12$ W/(m² K)			Flat roof Thermal transmittance $U_2 = 0.10$ W/(m² K)		
No.	Material	T (cm)	No.	Material	T (cm)
1	Silicate thin-layer plastering	–	1	Gravel 16/32	5.0
2	EPS (expanded polystyrene)	30.0	2	Secondary roofing (geotextile)	0.5
3	Outer masonry wall	25.0	3	Pitched XPS (extruded polystyrene)	5.0
4	Cement mortar	1.5	4	XPS	30.0
			5	Bitumen cardboard as waterproofing in two layers	1.0
			6	RC roof slab	20.0
			7	Finishes	–

Appendix: Examples of the Use of the Methodology for Evaluating Structural Details 201

Building Physics

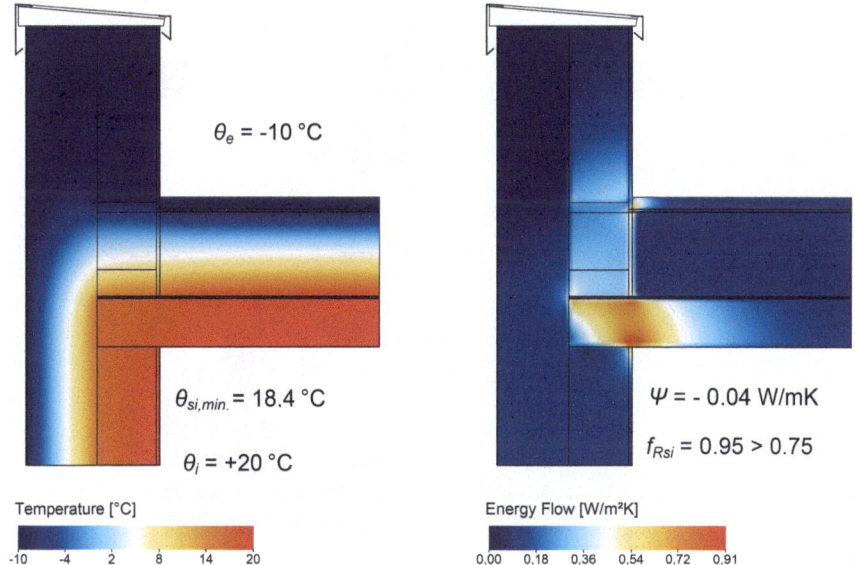

Detail Evaluation

		Score	Selected weighting factors (weights)	Corrected weighting factors (influence of external factors)	Share of weighted score
Environmental & energy-efficiency parameters	E1	6	1	1	0.86
	E2	6	1	1	0.86
	E3	6	1	1	0.86
	E4	6	1	1	0.86
	E5	6	1	1	0.86
	E6	3	1	1	0.43
	E7	5	1	1	0.71
Technical & structural parameters	K1	2	1	1	0.29
	K2	2	1	1	0.29
	K3	2	1	1	0.29
	K4	3	1	1	0.43
	K5	5	1	1	0.71
	K6	4	1	1	0.57
	K7	4	1	1	0.57

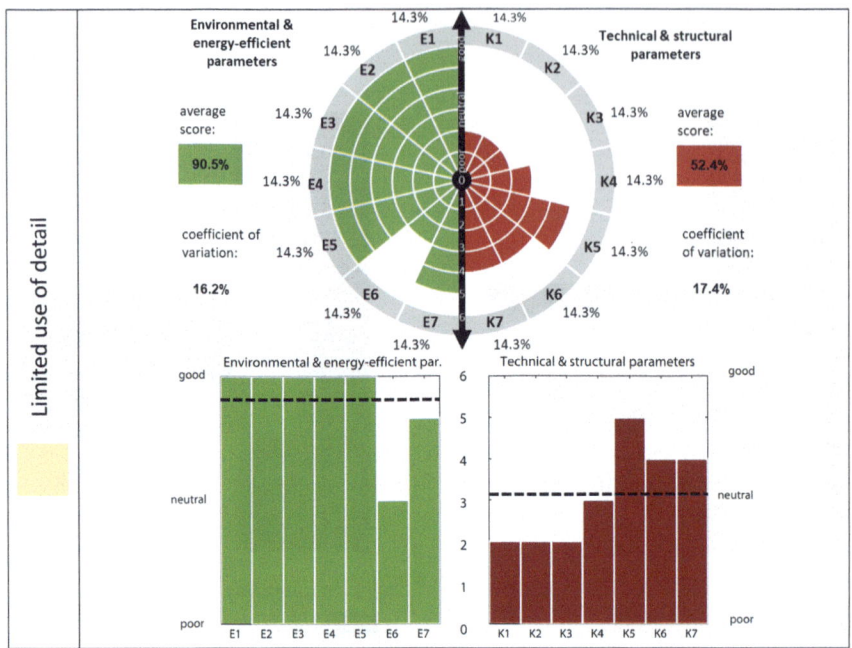

Concluding Remarks

Critical in the analysed contact is the fixing of the parapet on the base RC roof slab. The anchorage of the parapet suffices for static loads, but does not suffice for two-way horizontal seismic action. In higher parapets, seismic action can lead to a generally bending response, thus it is recommended to place anchorage reinforcement on both sides of the cross-section. Such fixing could be used for lower parapets, in which mainly internal shear forces develop. In addition, the insertion of thermal insulation reduces stiffness where the parapet is fixed. The total structural score is lower, since, in the worst case scenario, the parapet could collapse, putting people near the building at risk. On the opposite side of environmental and energy-efficiency parameters, the score of the detail is high with the insertion of a load-bearing thermal insulation elements solving the problem of a thermal bridge in the concrete parapet.

Appendix: Examples of the Use of the Methodology for Evaluating Structural Details

Detail code	DS-05
Brief description	Building connection detail between an RC roof and an outer masonry wall with a parapet

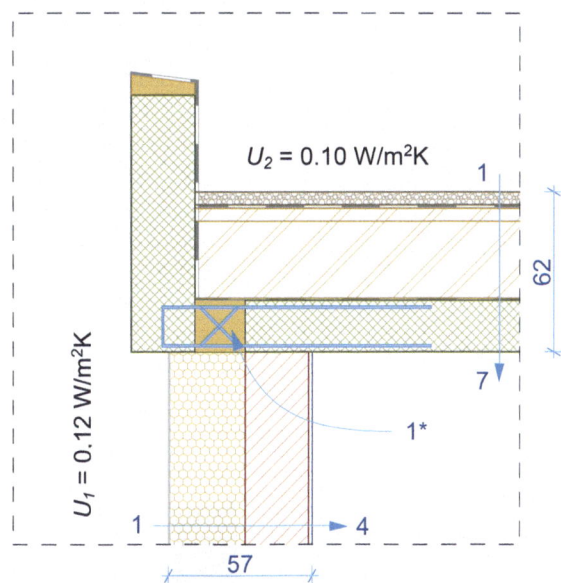

Detail Properties

The building connection detail between an RC flat roof and an outer masonry wall is analysed. The concrete parapet is displaced from the axis of the vertical load-bearing structure. The thermal insulation element (1*) with thermal conductivity λ = 0.12 W/(m K) (e.g. autoclaved aerated concrete, cellular glass, etc.) prevents a thermal bridge resulting from heat flow through the RC structure of the parapet. Steel reinforcement for the anchorage of the parapet is symmetrical.

The detail is composed of:

- outer masonry wall insulated with EPS ($U_1 = 0.12$ W/(m² K))
- RC roof slab insulated with XPS ($U_2 = 0.10$ W/(m² K)).

Structural Assemblies

Outer wall Thermal transmittance $U_1 = 0.12$ W/(m² K)			Flat roof Thermal transmittance $U_2 = 0.10$ W/(m² K)		
No.	Material	T (cm)	No.	Material	T (cm)
1	Silicate thin-layer plastering	–	1	Gravel 16/32	5.0
2	EPS (expanded polystyrene)	30.0	2	Secondary roofing (geotextile)	0.5
3	Outer masonry wall	25.0	3	Pitched XPS (extruded polystyrene)	5.0
4	Cement mortar	1.5	4	XPS	30.0
			5	Bitumen cardboard as waterproofing in two layers	1.0
			6	RC roof slab	20.0
			7	Finishes	–

Building Physics

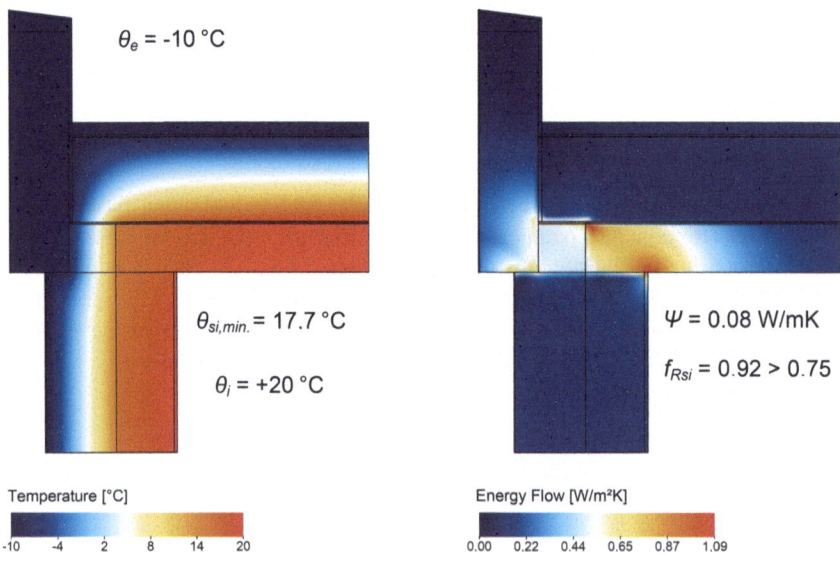

$\theta_e = -10\ °C$

$\theta_{si,min.} = 17.7\ °C$

$\theta_i = +20\ °C$

$\Psi = 0.08$ W/mK

$f_{Rsi} = 0.92 > 0.75$

Temperature [°C]
-10 -4 2 8 14 20

Energy Flow [W/m²K]
0.00 0.22 0.44 0.65 0.87 1.09

Appendix: Examples of the Use of the Methodology for Evaluating Structural Details 205

Detail Evaluation

		Score	Selected weighting factors (weights)	Corrected weighting factors (influence of external factors)	Share of weighted score
Environmental & energy-efficiency parameters	E1	6	1	1	0.86
	E2	5	1	1	0.71
	E3	6	1	1	0.86
	E4	6	1	1	0.86
	E5	6	1	1	0.86
	E6	3	1	1	0.43
	E7	5	1	1	0.71
Technical & structural parameters	K1	5	1	1	0.71
	K2	5	1	1	0.71
	K3	4	1	1	0.57
	K4	4	1	1	0.57
	K5	3	1	1	0.43
	K6	6	1	1	0.86
	K7	5	1	1	0.71

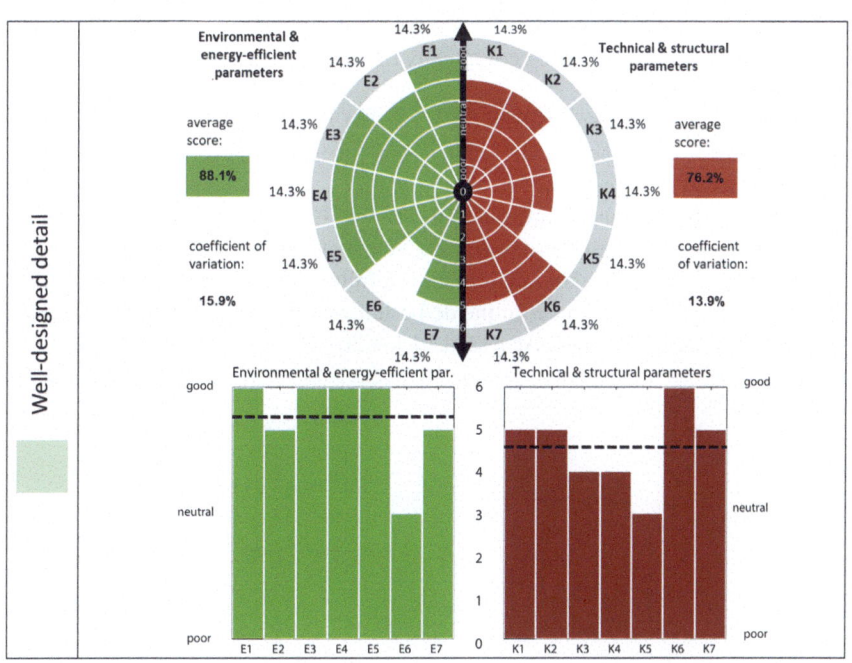

Concluding Remarks

A special feature of the analysed solution of a concrete parapet at the contact between the roof and an outer wall is that the axis of the vertical load-bearing structure is displaced in comparison with the placement of the parapet. This means that such a detail can be solved only with special precast elements with good load-bearing capacity and insulation properties. This also solves the problem of a thermal bridge in the concrete parapet, resulting in an excellent environmental and energy-efficiency score of the detail. On the other hand, the detail's technical and structural score is satisfactory with slight deduction due to the displacement of the parapet from the axis of the vertical load-bearing structure and a special precast detail. A good characteristic of the analysed contact is the fixing of the parapet on the base RC roof slab with symmetrical steel reinforcement, providing load-bearing capacity for a force couple, which stems from the bending moment brought on by the dynamic horizontal seismic loads, and load-bearing capacity for shear loads.

References

Azinović B, Koren D, Kilar V (2014) The seismic response of low-energy buildings founded on a thermal insulation layer-a parametric study. Eng Struct 81:398–411

Azinović B, Kilar V, Koren D (2015) Erdbebensicherheit vorgefertigter wärmegedämmter Stahlbeton-Konsolenelemente. Bauingenieur 90:489–499

Azinović B, Kilar V, Koren D (2016) Energy-efficient solution for the foundation of passive houses in earthquake-prone regions. Eng Struct 112:133–145

IBO (2008) Details for passive houses: a catalogue of ecologically rated constructions. A catalogue of ecologically rated constructions, 3rd edn. Springer, Vienna

Kilar V, Koren D, Bokan-Bosiljkov V (2014) Evaluation of the performance of extruded polystyrene boards-implications for their application in earthquake engineering. Polym Test 40:234–244

Schöck (2020) Technical information Schöck Isokorb® [Online]. Available: https://www.schoeck.com/en-gb/isokorb. Accessed 4 Dec 2020

Wienerberger (2016) Das Original Gefüllte Ziegel von Wienerberger [Online]. http://service.enev-online.de/bestellen/wzi_101221_gefuellte_ziegel_poroton-p_poroton-mw.pdf. Accessed 14 Jan 2016

The manufacturer's authorised representative in the EU is Springer Nature Customer Service Centre GmbH, Europaplatz 3, 69115 Heidelberg, Germany. If you have any concerns regarding our products, please contact ProductSafety@springernature.com

Printed and bound by CPI Group (UK) Ltd, Croydon, CR0 4YY

23/03/2026

02076661-0004